Home-grown Energy from Short-rotation Coppice

Home-grown Energy
from Short-rotation Coppice

George Macpherson FRAgS, FIMgt

Farming Press

ISBN 0 85236 289 7

A catalogue record for this book is available
from the British Library

Published by Farming Press Books
Wharfedale Road, Ipswich IP1 4LG, United Kingdom

Distributed in North America
by Diamond Farm Enterprises,
Box 537, Alexandria Bay, NY 13607, USA

Main cover photograph: Murray Carter, Chairman of British Biogen,
in a maturing crop of willow at Ingerthorpe Hall,
near Harrogate in North Yorkshire. (Clare Arron)
Other cover photographs by Murray Carter

Cover design by Andrew Thistlethwaite
Typeset by Galleon Typesetting, Ipswich
Printed and bound in Great Britain by Biddles Ltd,
Guildford and King's Lynn

Contents

There is a colour section between pages 152 and 153.

Acknowledgements

Writing this book has been a great pleasure. It has also been a great education and a fine way to get to know people in the new and exciting industry of home-grown energy. I would like to thank all those who have checked my chapters — and hope that they have not missed too many howlers. They have been very long-suffering and patient with my ignorance. I would like also to thank all those publishers and authors who have supplied reports, books and papers, some as yet unpublished, from which I have been able to glean useful additions to this book. Thanks, too, to all those who have supplied photographs and diagrams, charts and tables: their contribution has been most valuable.

I have tried to mention everyone concerned within the text, and if your name appears in the book it is most likely that I have been pestering you for information and advice — I am most grateful to you. If you have not been credited, please forgive me — it was not my intention to miss you out. My thanks to Jane, my wife and partner, for constructive criticism and practical help; and to my family for excusing my absence in the office. Finally, thanks to Roger Smith, the book manager at Farming Press, and to editor Julanne Arnold for their help and encouragement — and the opportunity to publish this book.

GEORGE MACPHERSON
Calne, Wiltshire

Foreword

DEREK WANLESS
Director and Chief Group Executive, National Westminster Bank

The need to move towards sustainable development and to achieve environmental goals raises important questions for the whole of society, whether these be at governmental, business or individual level. Of primary concern is the need to develop clean sources of renewable energy which enable the most effective use of land no longer required for food production.

Current research indicates that thousands of hectares of agricultural land will become surplus to requirements over the next decade or so. This will have a profound effect upon rural communities and their way of life, and if we are to mitigate the economic impact on their lifestyles it will be necessary to generate and develop new income streams for their use.

Throughout the world the demand for energy is increasing as many nations seek to industrialise and raise the living standards of their local populations. Fossil fuels are a limited resource and the production of energy from them can result in the emission of pollutants to the atmosphere which may damage the ozone layer and cause harm to both flora and fauna, including, of course, the human population.

Of the currently available renewable sources of energy, biomass appears to present the fewest problems. It produces controllable heat and power; it is not unsightly; it helps to absorb atmospheric carbon dioxide; it can be grown economically on large areas of land providing an excellent habitat for a wide range of wildlife; it thrives on suitably processed waste and it can result in the production of large volumes of energy.

Sweden and Britain lead the science and technology for biomass production in temperate climates. Both nations have invested over 15 years of research and development in this area, particularly that relating to willow and poplar trees grown as arable crops using agroforestry techniques and more popularly known as 'short-rotation coppicing'.

The recent announcement of contracts for the production of electricity under the Non Fossil Fuel Obligation and the growing awareness by society of the need to use sustainable forms of energy makes the publication of this book particularly timely. It provides policymakers, farmers, landowners, engineers, architects, planners and advisers with a practical guide to the understanding of short-rotation coppicing and its immense potential.

As Chairman of the Advisory Committee on Business and the Environment (ACBE), I am delighted to see British business taking the lead in such an important area, demonstrating clearly that environmental sense and business sense can go hand in hand.

Introduction

If you are a farmer or landowner and have not yet planted any willow or poplar — even simply on a trial basis — it is time to start. Here is an opportunity that is not to be missed, if, like farmers worldwide, you are worried about what to grow next for a profitable market. In Britain we harvest superb cereal crops and excellent grass but, despite burgeoning world population, the market for our grain and livestock produce hardly increases. The Third World cannot afford this expensive food. Quotas are being applied to many of our major products, and farmers in the European Union face increasing competition not only from former communist countries but also from the Americas and elsewhere as GATT begins to bite. The squeeze on margins is becoming intolerable.

Arable farmers face the most immediate threats, but livestock producers may also be caught by even more sudden change if political pressure to withdraw subsidies reaches a climax. We need a completely new sphere of production of crops for which there is a healthy and growing market. Niche markets are great for those who can find them, but what is needed is a massive market requiring huge areas of land for profitable production. Various non-food crops show promise — for example, oilseed rape for manufacturing plastics — while fibre and oil crops like linseed and evening primrose may fill some gaps; but, to the frustration and embarrassment of the farming community, set-aside is having to be used to soak up surplus land. Farmers are being forced by economic circumstances to accept government money to become 'custodians of the countryside', rather than being able to follow their instinct to produce something valuable, of which they can be proud, in a way acceptable to the general population. They need mainstream crops with public appeal which will bring in a sensible income.

Of all the possible alternatives for new crops to fill this role, short-rotation coppice for energy production looks the most promising. It is on the point of being adopted as a mainstream crop, with the prospect of covering many hundreds of thousands of hectares in the United Kingdom over the next few years. A huge amount of research and development has already been carried out in both Sweden and the UK. Short-rotation coppice gives farmers the opportunity to use government and European Union assistance in a constructive and productive way rather than simply leaving land under unsightly set-aside. In Britain we have the planting material, and we are well on the way to overcoming many problems of disease and weed control, harvesting and utilisation. Government has decided to give more incentives to the development of energy crops.

This book is for farmers, landowners, students of the land and water sciences, conservationists and anyone interested in actively pursuing a personal target of tending renewable global resources in such ways as to allow humanity to farm its way into a virtuous circle of production and consumption of food, fibre and energy. Some farmers will opt for production of woodchips as a commodity for sale to electricity generating companies, or for other purposes such as mulch or pulping, while others will choose to sell electricity or heat locally. Whatever the case, in ten years' time we shall take short-rotation arable coppice as much for granted as we take cereals or grass. Now is not the time to decide *whether* to plant, but where to plant, what to plant and how much; and, since the crop lasts up to 30 years in full production, it will pay to get things right.

Farmers, knowing the many risks involved with new crops, are likely to approach short-rotation coppice with caution. Much of this book is devoted to ways of doing this, but there's also plenty of guidance for the bold if they want to put more on what is undoubtedly a winner!

CHAPTER 1

The Market for Woodchips

There is a major new enterprise out there just waiting to be harnessed by farmers and landowners. It's called short-rotation coppice by some people and arable-energy coppice by others, but terms such as biomass, energy forestry and bio-coppicing are also used. It consists of poplar or willow trees, but to distinguish it from timber production it carries the 'short-rotation' label because it's harvested every three years, as compared with every 15 years with conventional coppice, and it lasts only 30 years in the ground, which is short in forestry terms. (Some forestry rotations reach 120 years.)

It is 'arable' because you prepare the land for it almost precisely as you would for wheat or barley — and you can rip it up easily to plant conventional crops if the need arises. This was proved at Long Ashton Research Station on the outskirts of Bristol when farmers expressed concern about 'losing' arable land. Coppiced willow was harvested and the roots then cultivated out using conventional equipment. Another way tried at Long Ashton was spraying off the new growth of willow in the first summer after harvest with glyphosate. The roots then died off and, as Rod Parfitt of Long Ashton put it, 'You can just kick out the stumps with your boot, after 18 months or so.'

It's 'coppice' because, at the end of the first year, the closely spaced trees are cut off near to the ground. Willow will send up ten to a dozen fast-growing shoots and poplar two to six stems, which, when harvested after periods of growth between two to four years, will produce an average of 10–18 tonnes of oven-dry wood per hectare per year (or 20–36 tonnes of 'green' wood).

3

Murray Carter inspects two single-clone plots of coppiced willow in spring. (Clare Arron)

As with any new crop, techniques of cultivation are changing rapidly and some growers are considering coppicing the crop after two years to obtain earlier productivity of usable material. It will take time to see which system works best.

As a crop, then, it is slow off the mark, taking some two to four years to become productive, but many farmers and land-owners can fill that production gap by cashing in on existing resources on the farm — using woodland by-products such as top and lop, brash and thinnings. One farmer, Rupert Burr from near Swindon, is bridging the gap by offering to clean up his neighbours' woodland *and* pay for what he takes until his willow coppice comes on stream.[1]

SURPLUS OF FARMLAND FOR FOOD

According to Professor Brynmor Green of Wye College, Britain could have as much as 6 million hectares of land surplus to our food requirements by the year 2000![2] He believes that out of the 127 million hectares of agricultural land in Europe as a whole, we probably need only 27 million hectares to feed ourselves. The professor is by no means alone in this belief, as John North of the Department of Land Economy at Cambridge has been giving similar advice for several years.

What has been in short supply, however, is energy without the disadvantage of extra carbon dioxide being poured into the atmosphere, without the possibility of nuclear fallout or the need for dangerous stores of nuclear waste, and without other polluting waste products. Short-rotation coppice has none of these disadvantages — it provides one of the purest fuels, and farmers could sell their products to produce heat, or electricity, or both. Other energy crops, such as perennial grasses like miscanthus or reed canary grass, and even annuals like sorghum or maize, may join in later.

Short-rotation coppice, for now, offers everything that the farmer could want — low labour, high production, low fertiliser or chemical requirements, large acreage, use of existing machinery (standard arable implements for seedbed preparation and almost standard forage harvesting gear for coppice harvesting), environmental enhancement and a high potential for public acceptance. But will anyone buy it?

SELL FIRST, PLANT LATER

Farmers have often been accused of growing a crop before knowing where it will be sold. In the case of energy, however, the markets are there before the crop. As a result of international pressure to reduce the quantity of carbon dioxide gas being poured into the atmosphere — which is suspected of creating global warming through what is known as the greenhouse effect — the British government has placed an obligation on companies producing electricity that they must supply a small percentage from non-fossil fuels. This rule is called the Non Fossil Fuel Obligation and is generally called NFFO (pronounced noffo). I will give more details about it later but suffice to say now that it is a 'pump-priming' initiative and it aims to encourage the production of electricity from a variety of sources — wind, wave power, hydroelectric schemes, landfill gas, municipal and industrial waste, farm and forest by-products and biomass — to help get new schemes launched.

Electricity companies are already mobilising their buyers to persuade farmers to take 10–15 year contracts to supply wood-chips. The South Western Electricity Board (SWEB) assisted by agricultural merchants Banks of Sandy, for example, has been holding farmers' meetings and putting forward plans for power stations in Cornwall, Hampshire, Suffolk and Northampton-shire. At Indian Queens in Cornwall, SWEB is planning a power station to produce 2.5 megawatts of electrical power — sufficient for 5000 homes. That would require 1500–2000 hectares of short-rotation coppice. At the time of writing it was set to be one of the first of its kind in the UK if it received a contract under the third NFFO arrangement.[3] A large number of groups are applying for contracts under NFFO — far too many for the available subsidy on electricity price — and results are to be announced in the late autumn of 1994.

A fourth NFFO is expected to be established for 1995/6. SWEB's other proposed sites would require about 3000 hectares of willow each. East Midlands Electricity is identifying power production projects for both willow and surplus straw. Southern Electricity has worked up a joint venture with Group Cereal Services, one of the UK's biggest grain co-operatives, to build a 20 megawatt power station to run on straw or woodchips. In the Scottish Borders farmer John Seed and associates have set up Border Biofuels, applying for a NFFO contract to produce electricity. The company will also build other generating plants combining heat and power production from 150 kilowatts to 5 megawatts using wood as fuel. The wood will come from forest by-products and short-rotation coppice. John Seed says that within 45 kilometres of Border Biofuels there are 12,600 hectares of set-aside land which could rise to 16,000 hectares of redundant arable land. If his plans come to fruition Border Biofuels will need more than 50,000 tonnes of green woodchips a year for a 5 megawatt plant. That would mean some 2000 hectares of coppice.

John Seed is not, however, asking farmers and landowners to plant hundreds of hectares each. Addressing a conference at the National Agricultural Centre in 1993 he said: 'I feel that it is unlikely that farmers will commit a significant area of arable land to this crop in the near future. For my own part, I would find it difficult to go beyond the 5 per cent level and in a survey including landowners and tenants, the response ranged from 1 per cent to 7 per cent of arable land.' He said that no one would plant a single hectare unless there was a positive margin guaranteed for an initial ten-year period.[4] Border Biofuels has targeted 1.5 per cent of the arable land in the Borders, representing some 1500 hectares, as an achievable target over the next four or five years. John expects more than 70 farmers to participate in producing woodchips for the power station. He commissioned the Scottish Agricultural College to carry out a feasibility study, and without any set-aside payments for arable coppice, the study suggested that a gross margin of around £320 per hectare would be easy to achieve. John Seed waited before planting willow or poplar, though, for the Ministry of Agriculture and Brussels to guarantee non-rotational set-aside for at least five years.

FARMER ENTREPRENEURS

When it comes to looking at returns from home-grown energy, Yorkshire willow importer and breeder Murray Carter, one of the pioneers of short- rotation coppice in Britain, prefers to use the term 'net margin' rather than 'gross margin'. Gross margin, he says, is gross output less variable costs. Better to take it one step further and deduct fixed costs to get 'net margin'. Gross margins do not give short-rotation coppice growers a fair picture of their returns: they suit budgeting for annual crops but not long-term crops like willow, which can stay in full production for at least 15 years and frequently for 20 or 30 years. This spreads establishment costs — and, once established, the fixed costs are very low.

At the other end of the country, in north Devon, farmer Alick Barnes, one of five farmers across the south of England selected to run demonstration plots of short-rotation coppice, has set up Green Fire Energy Ltd. He applied for a NFFO contract to set up a 600 kilowatt power station selling electricity direct into the national grid. An important by-product is the heat produced by the burning process. Alick is considering a range of applications for this, from heating his farmhouse and conference centre at Loyton Farm, Bampton, to heating plastic tunnels for horticulture or for a hydroponicum in which to grow tropical fruit or out-of-season strawberries, or for heating water for a fish farm or tourist attraction. Green Fire Energy Ltd is targeting slightly smaller electricity generation units than Border Biofuels. Alick's 600 kW plant will run with six 100 kW generators. This gives more security of supply, as output tends to fluctuate when using woodchips as fuel. He will need 600–700 hectares of coppice but initially will be using woodland residues and by-products as fuel. The company will be supplying small-scale gasifiers and consultancy on NFFO applications. Rupert Burr's project at Swindon is similar to that of Green Fire Energy, but is on a smaller scale.

STEP BY STEP

So far we have looked at two variations on the theme of growing short-rotation coppice for electricity production — the first, with farmers selling woodchips as a commodity through merchants to electricity companies; and second, with farmers themselves producing electricity and selling direct to electricity companies, delivered into the national grid. John Seed, Alick Barnes and Rupert Burr are adding value to their product — and offering to assist other farmers and landowners in setting up similar enterprises.

It is of interest to see what is happening in the United States of America regarding electricity production from wood. In the state of Maine, which is the size of England but with a population of just 1.3 million, most of the land is forested. The state consumes 2800–3000 megawatts of electricity annually. In 1981 less than 1 per cent of that power was supplied by biomass. By 1990 this had risen to over 20 per cent, and by 1993 there were 23 independently owned biomass power plants with outputs of more than 1 megawatt. There is still enough biomass available for wood-fired power plants to cope with another 600–1200 megawatts, according to American project developer Brian Chernack of KTI Energy Ltd.[5] He believes in larger schemes — over 10 megawatts — for long term success. He also believes it is a mistake to build a plant for only one type of biomass: there should be provision to burn wood, peat or straw, for example.

CAUTION

For UK farming families who depend almost entirely on their land for a living, a large-scale commitment to short-rotation coppice for energy may seem too ambitious. A survey carried out for the Energy Technology Support Unit (ETSU, pronounced ettsue), based at Harwell in Oxfordshire, by the Centre for Agricultural Strategy[6] showed that the farmers being studied would only consider planting between three and seven hectares of energy coppice on their land because they viewed large-scale planting as too risky. In the experience of Yorkshire farmer and plant supplier Murray Carter, however, farmers' initial caution about planting too much coppice is usually overcome by their enthusiasm for the crop after the first year or so. 'Once they've tried it they tend to plant larger areas each succeeding year, as their confidence grows,' he says. This suits Murray well, because as will be seen in the chapter on planting material, you don't plant one single clone or variety on your farm, but several clones, so as to help avoid any breakdown to diseases.

Bearing in mind the understandably cautious attitude of farmers, it might be best to approach the market for short-rotation coppice in two phases — first by selling heat, then electricity later on. Starting locally, a group of farmers might set out to identify potential markets for heat: for example, establishments which use high levels of heat, such as schools, factories, sports centres, hospitals, prisons or stately homes. They might form companies, co-operatives or partnerships to carry out feasibility studies for each project to see whether they can compete with existing fuels such as oil, natural gas or coal.

Investment by each farmer can be kept at acceptable risk levels, while everyone gets accustomed to the crop, the management of its utilisation and its problems, but sufficient short-rotation coppice can be planted in comparatively small areas over the required number of farms to supply not only the projected market but also the farmhouses and various farm enterprises on the farms. Woodchips will not need to be transported very far and forage trailers will probably have the necessary capacity. Obviously margins will be tight at the beginning, as it takes time for the crop to come into production, as mentioned earlier in this chapter.

After four or five years, when heat sales are running smoothly, a limited supply of electricity might be considered. Peak demands for current would still have to come from the national grid because the locally produced supply would only be capable of steady output. Selling heat alone, producers would not get the benefit of the higher price offered for electricity under the NFFO scheme — although at present a proportion of the establishment cost for the coppice could be carried either by the Woodland Grant Scheme or by non-rotational set-aside.

FIRST IN THE UK

Northern Ireland has the first combined heat and power unit in the UK, fuelled by willow, at the Northern Ireland Horticultural and Plant Breeding Station at Loughgall, County Armagh. Malcolm Dawson of the Department of Agriculture for Northern Ireland has spent many years on the project, installing a Belgian-designed gasifier powering a 100 kW generator at Enniskillen Agricultural College, with government and European financial assistance, and help from the college authorities, Mr G. Forbes and engineering consultants Power Management Associates (Bernard Wilkins and Christiane De Backer).[7] The unit supplies both electricity and heat to the

Ten acres (2.4 ha) of coppiced willow at Castlearchdale in County Fermanagh, Northern Ireland. (Department of Agriculture for Northern Ireland – DANI)

college from 16 hectares of short-rotation willow. The college may also contract local farmers to supply more willow, which would enable sales of surplus electricity to be made to the national grid.[8] Many lessons are being learned from this initiative.

Another exciting project is taking place at ADAS High Mowthorpe research centre at Duggleby in north Yorkshire, where Michael Green is working on a small scheme suitable for farm use. The centre is planting 50 hectares of willow over the next three years to fuel a gasifier generator to produce between 150 and 200 kilowatts of electricity. The heat generated by the gasifier and generating engine will be used for various purposes, such as heating buildings and glasshouses, or soil warming. The centre is also planning to install a research and development gasifier which will be able to produce gas from a range of farm-grown fuels, such as poplar and miscanthus, as well as from domestic waste and farm by-products. Michael says that power from this may even be used to charge batteries for electric farm vehicles.

Arthur Hacking is working on heat from willow at the ADAS Pwllpeiran centre at Aberystwyth, while at the ADAS Arthur Rickwood research centre at Ely in Cambridgeshire Colin Speller has been trying out miscanthus, reed canary grass, artichokes, thistles and sweet sorghum as possible farm-grown fuels for the future. Colin says that willow (and possibly poplar) are the only energy crops for which there is enough detailed information available to enable farmers to go ahead and invest.

Enniskillen Agricultural College. The gasifier buildings are in the foreground. (DANI)

HEATING A STATELY HOME

At the 1800 ha Drayton Estate near Kettering, in Northampton-shire, woodchips are being used to supply hot water and heat for a large stately home and three staff flats. Some 210 tonnes of air-dried woodchips are used per year (chipped by a tractor-mounted T70 Turner woodchipper). Fuel comes from estate woodland thinnings and by-products, offcuts from a local saw-mill and from short-rotation coppice. At Drayton the agent, Edwin de Lisle, and the consultant, John Lockhart (of Samuel Rose) knew there would not be enough wood available from the estate's woodlands alone so, initially, they planted a quarter of a hectare of willow and poplar trials, working with Ken Stott of Long Ashton Research Station. They planted 18 varieties of willow and 6 varieties of poplar, in 10 metre by 5 metre plots. In 1989 they extended the trial and planted another half-hectare with 9 of the best performing clones of willow and 3 of poplar. These trials proved most useful in learning about the crop and they are now poised to expand into large-scale plantations.[9] The installation of the boiler and total capital cost was £80,000, but the estate is saving £16,000 a year on electricity and oil. It used to spend £25,000 and now spends only £9000 a year.[10]

Another commodity market for the future, for farmers grow-ing short-rotation coppice for heat, may well be to replace straw in the straw-burning power stations with woodchips. At present, straw prices and availability make straw-burning power stations an economic proposition but if cereal growing diminishes, and straw use by livestock farmers increases, it may be necessary for those stations to turn to other forms of home-grown energy, and this could include woodchips. Such power stations are normally being built with the capability of using more than one type of fuel.

And just consider the glasshouse market. In the UK's tem-perate climate we can produce biomass very competitively so the cost of heating glasshouses could be significantly reduced, poss-ibly giving our horticultural industry a new lease of life, fol-lowing impossible competition from produce from the warmer climes of the Mediterranean.

NON-ENERGY MARKETS FOR ARABLE COPPICE

When planting willow or poplar coppice it is always comforting to know that there may well be other markets, apart from energy, for produce. There is a growing demand for charcoal, for example, so this is one area that needs further investigation. Apparently the size of wood most suited for charcoal production is 'wrist thickness' as compared with the usual 'thumb thickness' of the rods from arable coppicing, although charcoal burners used to build stacks of up to 5 tonnes of thin coppice wood which they covered with blankets of turf and set alight. After two or three days the fire died out and when the stack cooled it was opened to reveal about a tonne of charcoal.[11] It's certainly worth further investigating, because in the UK we produce only 2000 tonnes of charcoal but import something like 60,000 tonnes of it. Portable kilns might be a possibility: Geoff Buchan of Hadlow College, Kent, told an NFU/Friends of the Earth seminar in London that one person, working 'reasonably hard', could gross £100,000 a year producing charcoal from an investment of about £15,000.[12]

Poplar coppice of wrist thickness could be ideal for making charcoal. Lionel Hill with some of his produce. (George Macpherson)

Farmer and cuttings producer Chris Whinney, from North Molton in Devon, says a local chipboard factory may become an extra market for his produce. The factory is one of two owned by Caberboard, and they are currently using some 5000 tonnes of woodchips a week. Admittedly they have not yet used short-rotation coppice woodchips, but the timber administrator, John Ranger, tells me that it's simply a matter of finding the right resin to cope with the higher percentage of bark that would be found in woodchips. A rough idea of the price is £23–£24 a tonne of green timber. At a farmers' day at South Molton in Devon in September 1994 Caberboard talked about prices of £40 and more per tonne for dry woodchips delivered to their local factory. At Ebbw Vale in south Wales a new company, Tech-Board, is opening a £40 million factory with a capacity of 88,000 tonnes a year, making hardboard from home-grown wood, which should also give the market for woodchips a boost. Other possible markets include using woodchips as a mulch in parks and gardens, and for soft flooring at riding schools.

At present everyone has their own idea of what 'woodchips' means. To some it means dry lumpy sawdust, to others green 'mini-logettes' or 'fingerbits', while some people think wood-chips can only be produced by axes. The need for standards has been recognised by British Biogen, the trade organisation, formed in 1994 to guide this new industry to prosperity. Its motto is 'Power from the Land' and the head office is at the National Farmers Union headquarters in London. Membership includes the National Farmers Union, the Country Land-owners Association, Scottish National Farmers Union, the Association of Independent Electricity Producers, the Timber Growers Association, plant breeders, individual farmers and landowners, engineers, consultants and project managers. It liaises closely with the Department of Trade and Industry, the Ministry of Agriculture, the Department of the Environment, Friends of the Earth, Greenpeace, the Council for the Protection of Rural England and many others concerned with sustainable land use. British Biogen has appointed a working group to establish standards for woodchips, which can then be recom-mended for consideration as British Standards, and ETSU has issued a preliminary report prepared by consultant John Alexander of FEC Energy and Environment Services, Oldham.

One other use for wood produced from short-rotation coppice

Lionel Hill supplies
planting material for
'green sculptors' in
playgrounds . . .

. . . and for parks.

would be pulping. Wood pulp can be used in the manufacture
of paper and packing materials. Young willow and poplar do
have disadvantages for this purpose, such as their comparatively high content of bark, but I am told the market does exist.
Certainly pulp from other energy crops of the future, like miscanthus and reed canary grass, looks most promising for inclusion in high quality paper and packing materials.[13] More than 20
per cent of the overall United Kingdom balance of payments
deficit can be accounted for from the pulp and paper deficit,
while the 'all-fibre' deficit stretches the figure to over 30 per

cent.[14] In 1991 the UK had a total trade deficit of £13,909 million, of which 'all-fibres' constituted £4,189 million. Ian Morrison of the Scottish Crop Research Institute at Invergowrie told delegates to the 1994 PIRA conference at Silsoe that 'There are other sufficiently convincing reasons to pursue the search for novel sources of plant fibres to be used in conventional and novel applications. This search should not be done at the expense of wood fibres: both sources can easily be accommodated within the current demand.'

SUMMARY

There is, then, a market for short-rotation coppice. Indeed, there is more than one market, there are a range of markets, including electricity production, heat production, woodchips, wood pulp, mulching, playgrounds, horse arenas and charcoal. This is a very healthy prospect for farmers and landowners. The potential

These willow cuttings could end up as fuel, paper pulp, mulch, playground or arena safety surface or chipboard. They can be sold as a commodity, or can have value added before sale.

size of the market for heating and electricity is staggering! The European Union and the British government have pledged enormous support to turn this vision of sustainable energy production into reality. The farmers' unions, environmentalists and conservationists are beginning to give their approval. Now is the time for farmers and landowners to take the plunge — or at least, to dip a toe into the water.

References

1   Open day at Roves Farm, Severnhampton, Swindon, 27 January 1994.
2   Seminar on 'The non-food uses of crops and land, opportunities and obstacles', NFU/Friends of the Earth, 1993.
3   Growers meeting, 10 January 1994, at The Bull, Long Melford, Sudbury, Suffolk.
4   RASE conference on short-rotation coppice, 'Market development and power production in the Borders area', 1993.
5   Brian Chernack, in his paper 'Large-scale electricity generation from wood', given at the conference at Cambridge on 13–14 October 1993, 'Wood, a new business opportunity'.
6   ETSU report no. 1322, 'Farmers' current attitudes to energy forestry in Great Britain', 1993.
7   Malcolm Dawson, in his paper 'A small-scale gasifier-based combined heat and power system', given at the conference 'Wood, a new business opportunity' (see note 5).
8   Article in *Farmers Weekly* about Castle Archdale Experimental Husbandry Farm, part of Enniskillen College, 7 May 1993.
9   John Lockhart's paper 'Drayton Estate — the estate option', at the conference 'Wood, a new business opportunity' (see note 5).
10 ETSU leaflet 'Wood-fired heating and hot water: Drayton Estate, renewable energy case study'.
11 'Old ways, new uses', *Forest Life*, 1993, p. 5.
12 Geoff Buchan's paper 'Non-energy crops: wood', at the NFU/Friends of the Earth seminar, 'The non-food uses of crops and land: opportunities and obstacles', 1993.
13 'Non-wood fibres for industry', PIRA/Silsoe Research Institute joint conference, March 1994.
14 Ian Morrison, 'Range, provision and processing of non-wood fibres from temperate crops' PIRA/SRI conference, 23 March 1994.

CHAPTER 2

Why Energy Crops Could Fill More than a Million Hectares of UK Farmland

The production of non-food crops will expand very rapidly in the next five years, mainly because it is British government and European Union policy. There are many millions of pounds earmarked for the rapid development of home-grown energy. Governments are planning to avoid any open-ended commitment to support this new industry, but will, instead, provide pump-priming to get things moving. The 1990 Environment White Paper set a government target of 1000 megawatts of new, renewable electricity generating capacity to be up and running by the year 2000. Then, in 1993, the Coal Review boosted that figure to 1500 megawatts. To put this in context, a 20 megawatt straw-fuelled plant proposed for Calne in Wiltshire would produce enough electricity for a town of some 15,000 people and would require something like 150,000–180,000 tonnes of dry straw per year. I take straw as the example because that is the farm-grown energy fuel which will come on stream first. Woodchips from woodland thinnings and loppings will probably be next, while short-rotation coppice takes three or four years to come into production. The Calne plant, the result of a joint initiative by the UK's second-biggest grain co-operative, Group Cereal Services, and Southern Electricity would be built subject to winning a NFFO contract in the third round of the NFFO scheme — and, of course, subject to planning permission.

Short rotation's winter harvest in progress in Sweden, where home-grown energy has become a significant contributor to the national electricity grid and district heating systems. (Stig Ledin)

The design allows for woodchips to be used in conjunction with or as a substitute for straw.

In Sweden, where willow is already producing heat and electricity very reliably, a 2 megawatt district heating plant owned and run by a group of 20 farmers to serve 3–4000 people requires 500 hectares of coppice. This was planted in 1990 and is now coming up for harvesting. Ken Broad, of Oxfordshire County Council, who is raising the awareness of local woodland owners and farmers about the value of the county's many spinneys and areas of neglected woodland, has visited the operation in Sweden where he found one man running the computerised plant. During its first few years it ran on sawdust, bark and offcuts from local timber operations. Farmers top up the fuel supply each week and villagers pay for the heat they use plus a standing charge. The willow is fertilised with sewage sludge and the ash from the heating plant.

Sweden decided to switch from coal, oil and nuclear power to renewable resources in the mid-1980s. Nuclear power will be phased out altogether. Straw, peat, forest residues and energy coppice are already providing some 15 per cent of Sweden's total energy requirements. Farms there are small, at around

80–100 hectares, and constitute only 8 per cent of the land area. There are, as Ken Broad puts it 'swingeing environmental rules and regulations' in Sweden. According to the *Financial Times* Sweden already has 9000 hectares of energy coppice under commercial production.

There are plans for a wood-powered electricity generating plant to be built in Herefordshire by Pontrilas Timber. Given a NFFO licence, it will use forestry residues to generate 5 megawatts, serving the needs of 8500 homes. Only a small percentage of the government's target of 1500 megawatts by the turn of the century will come from short-rotation coppice and straw, but from the examples given above the impact that just a couple of megawatts can have on farming is obvious. That target of 1500 megawatts would represent only about 3 per cent of the nation's need for electricity, and coppice would only provide a very small part of that by 2000, but, according to the government-sponsored Renewable Energy Advisory Group, if only 16 per cent of agricultural land (and that's some 2.8 million hectares) were turned over to coppice by 2010, the energy produced from it could represent 22 per cent of the UK's electricity.[1]

At first most 'renewable' energy will come from existing and new hydroelectric generation plants, methane digesters, landfill sites and wind generators, but home-grown energy from farms and forests has advantages over all the others — at least so far — in that it can be produced when it is needed: to order. That's not possible with wind generation — the wind blows when it will! With home-grown energy it is possible to plan ahead, to produce more heat or power during cold periods, or at different times of the day. It can also be stored comparatively cheaply, for woodchips (if they are reasonably dry) or bales or bundles of willow rods can be stacked. To store solar power, or wind power, would probably entail batteries, while inefficient means such as pumping water uphill to reuse the energy would be needed in the case of water power. Bio-diesel produced from oilseed rape can, of course, be stored but the present GATT agreement limits industrial oil cropping to an EU-wide 1 million tonnes from set-aside land. Oilseed rape grown in the main scheme for arable area payments for industrial use counts towards the EU base area, which is set at 5.128 million hectares.

GOVERNMENT'S REASONS

The UK government, like the governments of all industrialised countries, is faced with expensive damage being caused to buildings, to human health and to the environment by emissions from the burning of fossil fuels. In a recent report from the Department of the Environment scientists say that in half of Britain — mainly in Wales, north-west England and south-west Scotland, with patches of East Anglia and around London also being affected — buildings and wildlife are being damaged by acid rain.[2]

Contraction of the coal industry will reduce the area affected by nearly nine-tenths early in the next century, while in March 1994 the UK agreed to the terms of a United Nations protocol to curb the sulphur dioxide emissions which help cause acid rain. These emissions come mainly from coal-burning power stations. Under the UN agreement the UK would have to cut emissions to half those of 1980 by 2000, to 30 per cent by 2005 and to 20 per cent by 2010.[3] The switch to natural gas will help the government to meet those obligations, and should also reduce damage to health, property and wildlife — but it does not address the problem of global warming, popularly known as the greenhouse effect.

THE GREENHOUSE EFFECT

'Natural' gas extracted from underground still releases huge quantities of carbon dioxide into the atmosphere, not to mention other gases such as oxides of nitrogen. Like most nations, the British government signed the Framework Convention on Climate Change at the United Nations Conference on Environment and Development in Rio de Janeiro in June 1992. The Convention is designed to control the emission of carbon dioxide into the atmosphere and reduce other pollutants, such as those which cause acid rain.

Governments are acting on acid rain because it is causing immediate damage and influential companies are beginning to find its effects expensive. Pressure groups too are having an impact. Damage from global warming is not yet so evident however, and since government policies are largely driven by industrial and commercial interests, little action is expected until phenomena like climate change begin to cost money. Scientists say that if we wait until hard evidence of global

Bio-diesel is easy to use and store but there are good reasons why it will only capture a niche market — perhaps for public transport in town centres.

warming is available, it will be too late to prevent many types of disaster, such as desertification, starvation and population migration. But how serious is this threat?

The greenhouse effect is said to be caused by the emission of carbon dioxide into the atmosphere. Every motor car, for example, gives out more than four times its own weight of carbon dioxide every year, according to Greenpeace.[4] Add to this the gas emitted by aircraft, from power stations, trains and ships, factories and homes and the figures reach frightening proportions. The greenhouse effect was first discovered in 1861 and since then, according to the United Nations Environment Programme and the International Union for the Conservation of Nature and Natural Resources, the quantity of carbon dioxide in the atmosphere has risen from around 270 parts per million to about 350 ppm. The world's leading climatic scientists are concerned that the rate of global warming is about to reach danger levels. By the year 2065, if emissions continue to be as high as they are now, the concentration is predicted to reach 600 ppm.[5] The results of that would be massive global flooding as the world's ice caps melted.

Signs of damage from global warming are certainly beginning to appear. Coral reefs are going pale and dying in the Caribbean Sea and the Pacific and Indian Oceans.[6] Hurricanes of unprecedented violence are seen in the Caribbean and cyclones of record strength are hitting Pacific islands; unusually fierce droughts have been experienced in parts of Europe, the USA and Africa. Glaciers have been melting at a remarkable rate. Ice cores taken from glaciers in China, Russia and Peru showed that between 1937 and 1987 temperatures were higher than for any 50-year period in the last 12,000 years. Between 1978 and 1987 the Arctic ice cap shrank by 2 per cent. Sea levels have been rising by 1 or 2 mm every year for decades. We are warned that if carbon dioxide emissions continue at the present rate, the global average temperature will rise by some 3 degrees Celsius before the end of the twenty-first century. 'This rate of warming is breathtakingly rapid,' says the environmental group, Greenpeace, 'many times faster than at any time in human history.' Greenpeace believes that such global warming will lead to many plant and animal species, both on land and at sea, becoming extinct. 'To stabilise greenhouse gas concentrations at a level which would not induce climate change means a fundamental change in the way the world uses energy', it says. 'We must move away from economies based on oil, coal and gas to those based on renewable energy.'

THE CLEFT STICK

Another reason why energy crops, in particular short-rotation coppice, will expand at a great rate is that farmers in the UK have very few alternatives. At present there are fibre crops and specialist oil crops; and there may be bio-diesel, if activists succeed with it. Theoretically, fodder beet could supply all the UK's needs for motor car fuel and it would take only 600,000 hectares to do so.[7] (One hectare of fodder beet can produce 60 tonnes of carbohydrate, which would convert into about 28 tonnes of liquid fuel (approximately 40,000 litres).) In 1993 630,000 hectares of arable land were taken out of production under the set-aside scheme. Perhaps fodder beet could be used as an annual energy break crop in arable rotations?

Large-scale food producers, though, will become few and far between. Farmers in Britain and Northern Ireland have been experiencing the most terrifying changes, as farmland used for food production is contracting at an unprecedented rate. The situation has been so traumatic that, during the past few years, the suicide rate in the farming community has been second only to that of vets. Immediately after the Second World War and until the early 1980s farmers were praised and rewarded for producing more and more food, of an increasingly high quality. Then, dramatically, in the mid-1980s, the brakes were applied, with quotas on milk production, quotas on sheep and set-aside schemes for arable crops. Let us consider some of the figures, supplied by the NFU.

UK AGRICULTURE TODAY

Agriculture is the UK's second largest industry, employing more than 600,000 people. British farmers grow 75 per cent of our temperate needs. At present we are the sixth largest wheat exporter in the world and the second biggest sheepmeat exporter.[8] But between 1991 and 1993 the total area under crops fell by nearly 10 per cent, to 4.5 million hectares. The area under wheat fell by nearly 15 per cent, barley by over 10 per cent and oilseed rape by 11 per cent; potatoes fell by over 7 per cent and horticulture by 8 per cent. (Horticultural income, incidentally, fell by a catastrophic 60 per cent in 1993/4.)

This is nothing, though, compared with what has to be faced over the next few years. It depends who you listen to, about just how little land will be needed in this country in six to ten years' time, to provide us with all the temperate food we require. In the last chapter I quoted Professor Bryn Green as saying that, here in the UK, we may have as much as 6 million hectares surplus to food production needs in just six years' time. Addressing the NFU/Friends of the Earth seminar in London just before Christmas 1993, he said 'we should not be surprised' at these figures. 'Two-thirds of the agricultural area in Britain is used for growing high quality protein. That is the process of using grass in producing meat. But two-thirds of the arable land in Britain is also producing cereals that are fed to livestock to produce high quality protein. At every step in the food chain about 90 per cent of the energy is lost, so there is enormous potential for more efficiency in agriculture'.

Apparently, then, the way in which we have always produced food is so inefficient that there must be this huge potential to grow enough food on far less land. Professor Green says that, even if you take a very moderate rate of improvement in efficiency — say 1.5 per cent per year — then 'with levels of self-sufficiency from 75–90 per cent; and demand little changed from the present (because the population is more or less stable in this country, indeed in north-west Europe as a whole, and we cannot eat much more because we are probably already eating far more than is good for us)' — we can extrapolate that, bearing in mind that we will eat as much as we can, procreate as much as we can and try to produce 90 per cent of our own food, there may well be a million hectares of land surplus to food production in six years' time.

'This is an awful lot of land,' says Bryn Green. 'It is equal to all the land that has been planted by the Forestry Commission, for example, since 1919.' Assuming that things go on more or less as they are now, with an increase in production of 2. 5 per cent a year and 75 per cent self-sufficiency — then there is a 6 million hectare land surplus. And that surplus doesn't take into consideration competition from outside the UK — it's related to improvements in our own production, efficiency and consumption. However, two days' lorry ride from the UK via the Channel tunnel, production of a new and copious supply of food is gathering momentum. Former communist countries are beginning to organise their farming businesses better. Professor Stephan Tangermann, a leading European agricultural economist, predicts that in only five years' time central Europe (including countries like Hungary, the former East Germany, Poland, Romania, the Czech and Slovak Republics) could have an exportable grain surplus of about 10 million tonnes, a sugar surplus of 2 million tonnes, a potential for net exports of 800,000 tonnes of pork and a further 800,000 tonnes of beef, and a butter surplus of some 400,000 tonnes.[9]

Once the EU expands to welcome such new members, any of our supermarkets could set up packing stations in central Europe (some are indeed doing so already), where production costs are minimal compared with ours (at least for the present). UK food producers would then face irresistible competition. UK farmers are aware of this threat too, and have long been racking their brains about what they could do instead of producing food. 'Diversify,' said the advisors — and those who could, did. But many of the changes have not produced sufficient income to substitute for the turnover being lost, the lower prices for produce and the higher costs of fertilisers, seeds, wages, agro-chemicals and rents. There is only so much bed and break-fast, horsiculture, clay-pigeon shooting, farm parks and evening primrose oil that the market can stand. Pick-your-own, adopt-a-sheep, golf courses, eventing courses, campsites, nudist colonies and battle-games: British farmers have diversified into all of them, at great expense and, in a few cases, very profitably. There are many government schemes such as Environmentally Sensitive Areas and Nitrate Sensitive Areas which give grants to farmers and landowners for *not* doing things, or for reverting to less productive methods. Apart from learning to play the game by the new rules, there is little excitement or potential for expansion and development in any of this.

So far, quotas have been good news for dairy farmers, creating a local shortage of milk and forcing prices higher — but for how long? If quotas persist, UK producers will fall behind and lose any competitive edge they ever had. In the long term, quotas of any kind are bad news and farmers realise this: quotas take them too far from the reality of the market place.

During the past 15 years, at the same time that the brakes were being applied to UK farming, many new pressures began to come from environmentalists and the conservation lobby. During the great farming expansion, the god of 'more cheap food' took precedence over good land use. Most farmers would (eventually) admit that widespread damage was caused to the countryside and to wildlife. Pollution and contamination went largely unchecked for many years. Then farmers were brought up with a jolt as more of them faced prosecution, and in some cases imprisonment, for contravening laws which were now being applied more diligently. Inexorably, management freedom has been restricted by: a shrinking market; rising costs; more regulations and more controls; and frequent attacks in the media and sometimes even face to face from the green lobby. All these negative factors have been confronting an ageing farming community — and just at a time when farmers had felt they were looking forward to an era of prosperity.

To cap it all, taxpayers, increasingly reluctant to pay huge subsidies to the industry, have put pressure on politicians to keep a firm check on all the money being handed out. This has resulted in the complicated IACS forms, farm maps and rigid discipline that went with them. Many farmers have been unable to face the changes. Some, still with enough equity to be able to retire, did so; but many have said publicly that 'I'll continue farming until I can no longer afford to', running up debts that ensured complete bankruptcy when the creditors finally closed in. There was no ray of hope, until the concept of non-food crops, such as fibre crops, oil crops and energy crops, began to catch some farmers' imaginations. For energy crops, particularly, there has been a most tremendous boost because of the government's introduction of the Non Fossil Fuel Obligation. Admittedly, the British government has been reluctant, to say the least, about adopting the European Union's proposal for a fossil fuel tax, or carbon tax, which has made home-grown energy instantly highly viable in countries like Denmark and Sweden — but NFFO is at least a start.

Farmers, then, if they can only see it, have a window through which they could emerge from much of the stultifying restriction that has engulfed them over the past ten years. Energy crops could change their lives for the better. The markets are there. The possibilities are endless!

But why grow willow rather than, say, whole-crop cereals? I put this question to Damian Culshaw, another of the specialists at ETSU, and he went through the pros and cons. He explained that grain has some attractions. It has about the same energy value as most biomass, something like 19 gigajoules (GJ) per tonne when it's oven dry. Freshly cut coppiced willow has at least 50 per cent moisture, whereas it's not difficult to harvest grain and straw at around 15 per cent moisture. That means a reduced energy value for the whole crop of about 15 GJ per tonne. Oil burners are easy to convert to grain — at least on a domestic scale. In Denmark there's a fossil fuel tax which makes coal, gas and oil much more expensive, so grain became competitive with oil for district heating or domestic heating. However, there's legislation to prevent grain being used for district heating, for political reasons: it obviously wouldn't do to burn grain when millions are starving elsewhere in the world. And although winter wheat can produce high yields of biomass, it needs planting every year, and it also needs high inputs of fertilisers and sprays to obtain the 19 or 20 tonnes per hectare at 85 per cent dry matter. A more realistic dry matter yield is about 12 tonnes per hectare. Short-rotation coppiced trees or perennial grasses such as miscanthus can easily compete with that.

In the UK there's no fossil fuel tax yet (except the VAT being charged on heating fuel), but grain cannot compete with coppice, because of the high cost of producing cereals. There are other arguments against the use of grain: it is, for example, very easy to add value to grain — by making it into bread, cake or alcohol — but it's expensive to grow and in terms of 'energy balance' gives poor results when used for fuel. There's a much better return using cereals for food. Grown conventionally, straw yields some 5 tonnes a hectare and grain about 8 tonnes, making a total of 13 tonnes, at 15 per cent moisture. That's approximately 12.5 dry tonnes. Coppice yields about 24 tonnes of green wood at 50 per cent moisture, which is 12 tonnes of dry matter — so almost as much fuel, but with far, far less by way of inputs, because coppice needs only about one-fifth of the fertiliser required by cereals.

Damian Culshaw says the energy balance is also in favour of coppice since it is perennial, lasting for up to 30 years. Whole-crop cereals are, however, easier to slot into existing arable systems in the short term, if a low cost, low energy system of establishment is used.

CHECKING UP ON THE FARMERS

Keeping a check on farmers and how they use their set-aside land gives bureaucrats headaches. They are adopting very stringent policing, with IACS (Integrated Administration and Control System) forms and even satellite imaging, to ensure that no food crops are grown on set-aside. Imagine having to differentiate between whole-crop cereals for energy and cereals for grain: willow or poplar coppice is much easier to distinguish!

References

1   A. Maitland, 'Trees branch out', *Financial Times*, 9 February 1994.
2   'Critical loads of acidity in the UK,' Air Quality Division, 1994.
3   B. Maddox, 'Half of country is damaged by acid rain', *Financial Times*, 11 March 1994.
4   Global warming campaign, ref. GPX W94, Greenpeace, 1994.
5   L. Durrell, *'State of the Ark'*. Bodley Head p 31.
6   See note 4.
7   J. Emsley, 'Energy and fuels', *New Scientist, Inside Science*, no. 68, 15 January 1994.
8   *Central Region Farmer*, February 1994, p.39.
9   A. Maitland, 'EU offers hope to Eastern Europe's farmers', *Financial Times*, 8 March 1994.

CHAPTER 3

Suitable Situations, Soils and Climates for Short-rotation Coppice

Short-rotation coppice should be planted as close as possible to where it will be used, because carting woodchips is expensive. It must also be accessible for harvesting in winter time. Getting bogged down with a large forage harvester and forage trailers full of woodchip in February's mud is not a pleasant thought. When the crop was first being considered for energy production, during the fuel crises of the early 1970s, even enthusiasts thought it only suitable for marginal land rather than prime arable sites. Now, however, with the prospect of so much arable and grassland becoming redundant for food production, some of the best land and sites can also be considered. But where does short-rotation coppice really thrive?

If you ask the research workers, 'Where's the best place to plant short-rotation coppice?' they are quite likely to say, 'We haven't a clue!' That's what Rod Parfitt of Long Ashton Research Station, speaking as a scientist, told me. He had to say that because, as yet, no 'proper' trials have been done, in his opinion. 'You would need small plots all over the British Isles, all planted in the same week or two, before a true comparison could be made,' he said. This is because the weather, soil temperature and soil moisture at the time of planting, and during the next few months, can affect a crop of short-rotation coppice for many years. Planting followed by a dry spell can handicap a crop, in terms of yield, so that it could take 10 or 15

Soil Type	Type of Clone
Amorphous, anaerobic clay and	*Willow*
light, loam soil	Most clones of S. × hirtei
	„ S. × sericans
	„ S. viminalis
	S. bebbiana
	S. burjatica 'Germany'
	S. × calodendron
	S. × dasyclados (Wimmer.)
	S. × fruitocosa
	S. × mollissima undulata hybrids
	S. stipularus
	Poplar
	× interamericana (though not as good as willows)
Silty, clay soils	*Willow*
	S. burjatica
	S. × dasyclados (Wimmer.)
	S. × mollissima undulata hybrids
	S. stipularus
	Most clones of S. × fruitocosa
	„ S. × hirtei
	„ S. × sericans
	„ S. viminalis
	Poplar
	× interamericana hybrids (though not as good as willows)
Drought-prone sites	*Willow*
	Avoid S. burjatica
	S. candida
	S. × hirtei
Alkaline	*Willow*
	Avoid S. burjatica
	S. candida
Light, but fertile soil	*Willow*
	S. viminalis
Wetter, more acidic soils	*Willow*
	S. aurita
	S. burjatica
	S. caprea
	S. cinerea
	S. viminalis

Some clones perform better than others in different soils and climates. The best advice is to establish a small plot containing as wide a range of poplar and willow clones as possible and observe them. (This information is taken from the Department of Trade and Industry's Agriculture and Forestry fact sheet, *Short-Rotation Coppice*, no. 3, dated April 1994. Further information from ETSU; see Chapter 15 for the address.)

years to catch up others which did not have that disadvantage.

Speaking, though, as a very experienced grower of willow, in particular — he is curator of the national collection of willow at Long Ashton — Rod Parfitt says that, as a general rule, willow will grow anywhere except in very shallow soils over rock, and in high lime conditions. He quotes thin soils on the South Downs as not being ideal but suggests that the 'iron bar test' is a good guide. 'Take an iron bar and see how deep the soil is; willow roots have not been found below one metre but the better the soil, the better the crop is likely to be. A shallow soil cannot be expected to give such good yields,' he says.

Land which is inherently fertile but which has a drainage problem can also be used for willow — although, before considering the use of this type of land, it is essential that farmers and landowners consult local Farm and Wildlife Advisory Group specialists because the land may be much more valuable as a conservation site. Impermeable clays like the gley soils or culm-clay soils of Devon and Northern Ireland are favoured by willow. 'Any land with a drainage problem, even on acid soils with a pH of 4–5 (but not above pH 6–7) will grow willow,' says Rod. The effect of too high a pH is iron chlorosis, in which leaves go yellow and brown and plants are stunted. Some clones of willow, for example *Salix viminalis*, can tolerate higher pH better than others, although Rod Parfitt says there is no empirical evidence to support this opinion. 'It's our experience rather than experiments,' he says. *Salix burjatica* is happier at lower pHs.

'The Swedes collected many species and clones of willow from across the tundras of Russia and northern Europe to use in their breeding programmes, which are now so advanced that they have clones to suit many arable soils,' says Rod Parfitt. Sites which are usually considered to be marginal for most arable crops, for instance because of excessive winter waterlogging, are often suitable for coppice. In Northern Ireland alone at least 200,000 ha of such land has been identified. Rod Parfitt and others originally suggested that a rule of thumb for selecting short-rotation coppice sites might be 'those areas of a farm currently agriculturally less productive, which would be the first choice for permanent set-aside' (except steep land), although now that the extent of surplus prime farmland becoming redundant over the next few years is becoming clearer, using higher quality land will be increasingly attractive as farmers pursue income and profit. Not only can coppice exploit conditions unsuitable for arable crops, it thrives on better land too.

Even in 'the early days' of considering short-rotation coppice crops for energy, large areas of land were considered suitable. For example in a leaflet, 'Willow biomass as a source of fuel,' published by Long Ashton in 1988/9, George McElroy and Malcolm Dawson writing from Loughgall and Ken Stott and Rod Parfitt at Long Ashton suggested that: 'Willows can grow satisfactorily on land which is often too wet for other crops and is therefore under-utilised. In England and Wales areas have been located and quantified by reference to the Soil Water Regional Map, and the National Soil Map of the Soil Survey.' The leaflet said that some 855,600 ha (8 per cent of the land area) had possibilities. Looking at soil maps and consulting ADAS suggested that 25,000 ha (5000 in Somerset alone) 'have real potential for commercial willow biomass production.' It went on to say that, in Northern Ireland, about 15 per cent of the province would be suitable. That was before the realisation that

Short-rotation willow coppice harvest in Sweden. (Damian Culshaw, ETSU)

up to a fifth of the UK's best farmland — not just the marginal land — would be redundant for food production before the millennium.

PIONEERS

It was in Northern Ireland that the idea of willow as an arable crop took root. I am reliably informed that it was George McElroy, in search of alternative crops for Northern Ireland's farmers, who took action to assess willow's possibilities as a commercial crop on 'disadvantaged land'. At the time, in 1973, there were projected world shortages of pulp from conventional forestry. Soon after McElroy's work began a series of fuel crises stimulated governments worldwide to seek alternative fuel sources, adding importance to the trial plots of willow at Loughgall and Long Ashton where the crop was also being looked at for horticultural shelter belts.

Until now only *Salix* (willow) and *Populus* (poplar) have given the consistently high yields and resilience needed for frequent coppicing, although other tree genera have been looked at. In the early days alder was tried, along with willow and poplar, but although it grew well initially it apparently did not coppice well. Eucalyptus, too, was considered, but it suffered from winter cold and infection of cut stumps by silver leaf disease. It would appear that poplar is more drought tolerant than willow, although both enjoy a good moisture supply. Peaty soils, such as are found in the fens, may not prove suitable due to problems with weed control: there is a tendency for the soil to make residual herbicides less effective. More research is needed to discover how to solve this.

If in doubt about where willow will thrive, look around the local countryside for willow trees. Consult local botanists and foresters. One of the pioneers of coppiced willow in this country is Ken Stott of Long Ashton Research Station. Now officially retired, he runs a consultancy from Long Ashton. He advises avoiding dry limestone areas and dry chalky zones. 'That would be difficult for willow, although poplars may be satisfactory,' he told me. Even in the Cotswolds willow might thrive: 'Just look at the size of other trees that grow there — you must choose your site.' Willows, however, says Ken Stott, are

One of Ken Stott's trial plots. One of the 'willow pioneers' in the UK, Ken is still based at Long Ashton near Bristol and works as a private consultant.

more hardy than poplar and grow further north than any other type of tree. In terms of rainfall, willow needs 500 mm or more per year, which is similar to the amount required by sugar beet or winter wheat. Coppiced crops can take water from up to one metre deep, again similar to sugar beet and oilseed rape although not as deep as winter wheat. Shrub willows, like those used for coppice, have shallow roots, as mentioned earlier. To sum up, short-rotation coppice and conventional arable crops need similar rainfall or soil moisture.

BUFFER ZONES

An interesting difference between willows and poplars is pointed out by Dr Nicholas Haycock of Silsoe College (part of Cranfield University). The poplar can have an extremely powerful tap root that may go as deep as 28 metres and can use a very large amount of water. 'It takes a lot of water from a stream, for example,' says Dr Haycock. In a situation where grassland allows the water table to stay at around 1.5 metres, planting poplar can lower that water table by 14 metres. This does not appear to occur with willow.

Dr Haycock's interest in willow and poplar is in their use in the prevention of agricultural pollution. Planted as 'buffer zones' next to rivers and streams they can help prevent nitrates and phosphates from farmland entering waterways. On deep alluvial soils the poplar can produce a root barrier between the arable land and a river. Even during the winter root growth and associated bacterial activity continue, so the poplar carries on its work as a 'pollution sentinel'. On less permeable soils the willow can do the same job.[1] Periodic harvesting of buffer zones is ideal, according to Dr Haycock, because it takes away nutrients that have been collected there. With poplar and willow regrowth further quantities of nutrients are taken up by the remaining root stools.

Although buffer zones need not be much more than 10–20 metres wide, the concept is very interesting for short-rotation energy coppicing. There is no reason why such a zone should prevent people walking along by the river — sufficient room can be left for a path and, indeed, it is good practice to avoid cultivation near river banks. The zones make a superb habitat for birdlife and small mammals but, as we shall see later in this book, it's a good idea to plan 'rides' or access routes into the crop so that where a view would otherwise be hindered, it is left open. A gap of a few metres can make all the difference between keeping or spoiling a treasured view (and between complaints and friendly co-operation with residents and the public). Long belts of coppice, then, in an appropriate place, will not only serve as profitable crops but also as protection against prosecution for inadvertent pollution.

Canadian research[2] indicates that long, rectangularly shaped plantings of coppice are preferable because they cut down the amount of turning and manoeuvring of harvesting equipment and, perhaps more importantly, they may reduce the amount of headland. (It is the headlands which get chewed up during harvest by tractors and trailers.) The coppice land itself is much better supported by crop roots. Long strips will, however, increase the cost of fencing, which is often necessary for the first year or 18 months to protect the crop from rabbits or deer: this needs to be considered and a balance sought. However, in conditions where soil and nutrient erosion is a problem, long belts of coppice can help alleviate matters.[3]

The Forestry Commission has issued some excellent advice on the layout of short-rotation coppice in its leaflet 'Short-rotation coppice in the landscape' (Forestry Practice Advice Note no. 1, obtainable from the Commission in Edinburgh — see Chapter 15 for the address).

THE LIE OF THE LAND

Short-rotation coppice is harvested in winter, so when choosing a site good access is essential. Steep slopes are not likely to make the job easy! This is a crop to replace food crops so if a combine can travel on this land, or if a forage harvester can work satisfactorily, coppice is more likely to succeed. Preparation for the crop will require deep cultivation and good seedbed conditions before planting and this also needs to be remembered when choosing the site for coppice. Both poplar and willow have a tendency to find drainage pipes with their roots — and to block them, especially newly laid pipes. Even though coppice is not difficult to remove, cleaning up the drains is impossible, although with willow there should be no problem if the drains are below a metre deep. In some parts of the UK the blockage of drains may be reason enough to preclude cropping with short-rotation coppice. In choosing your site for coppice, it's a good idea to consult the supplier of your planting material, who may advise doing tests to ascertain both the pH and nutrient status of the soil and should certainly be able to recommend the appropriate mixture of clonal types. When dealing with the electricity generating companies there should also be technical advice available on this question.

The chosen area should be planned into a number of equal plots: if working to a three-year harvesting cycle, three plots or strips, and so on. Leave between 10 and 20 per cent of the land unplanted, to allow for access, vehicle turning and possibly storage of cut rods or rod bales. Planting spaces and row width will be discussed in Chapter 6.

Planning and layout of the unplanted areas needs very careful thought, because the crop is going to be in the ground for up to 30 years, so far more than just the farming considerations need to be taken into account. The access rides, for example, provide very useful feeding grounds for birds of prey, or for sowing of species-rich wildflower mixtures (perhaps in conjunction with local conservation groups, although they must be prepared to see the land disturbed during harvest times). Such rides may well fit in with private horse-riding trails or nature walks. And, as mentioned earlier, some rides may need to be sited opposite houses, rather than spoil a valued view for local residents.

If the chosen site tends to dry out periodically, it may be necessary to irrigate the coppice during the first year of planting to ensure strong establishment. Don't forget that scrimping at this stage will be remembered by the crop for many years! The key to profitability in short-rotation coppice is high yield from carefully managed inputs. The better the site, the more likelihood of higher profits.

References

1 Interview with Dr Nicholas Haycock on BBC Farming Today, Radio Four, 25 April 1994.
2 J. Balatinecz, Handbook on how to grow short-rotation forests, Swedish University of Agricultural Sciences, Uppsala, 1992.
3 R. Parfitt, Graham Hunt, Edwin Thompson and Tracy Badmington, 'Wood production from short-rotation coppice, Mercia programme,' ETSU, 1994.

CHAPTER 4

Will It Pay?

Short-rotation coppice — wonderful opportunity; environmentally perfect; exactly what we all need — but does it pay? Will it pay? How much, and when? The answer is that it might pay but not easily and not very much, yet. Looking at the costs and benefits of home-grown energy is, as for any farming initiative, daunting, time-consuming and at times depressing, but there is sufficient hope of profit to keep going. As usual with farming, nearly everything will depend on the politicians. Politicians say they want 200 megawatts from energy crops and forest residues by the year 2000. This could happen if farmers and landowners were given sufficient pump-priming support by way of NFFO, combined with set-aside payments and woodland planting grants — and very easily if a carbon tax was put on fossil fuels.

The Department of Trade and Industry (DTI), the Ministry of Agriculture (MAFF) and the Department of the Environment (DOE) are working on a strategy for bio-fuels for the next ten years. The DTI is expected to devote some £1–£2 million a year on support activities for coppice cropping. In the European Union's budget there are huge sums of money earmarked for the development of renewable energy — something like 100 million ecu per year to be spent in the UK alone — although that will cover all forms of renewables such as wind power, wave power and tidal power.

Like all business opportunities, however, there are situations even now where woodchips can make very good money without government help. A farmer's eye could reflect the glint of the sun on nearby glasshouses, especially if they are heated by

40

liquid petroleum gas (LPG). Heat may be costing the flower or tomato grower as much as £7 per gigajoule. A farmer or land-owner should easily be able to produce woodchips for £2 a gigajoule, allowing him or her to install an appropriate plant to sell heat to the glasshouse business at an attractive rate all round. Government is looking at the possibility of a NFFO scheme for heat as well as electricity. This would greatly widen the potential for viable home-grown energy schemes in the heating sector.

Energy facts and figures. (This information comes from the DTI's Agriculture and Forestry fact sheet, *Short-Rotation Coppice*, no. 6, available from ETSU; see Chapter 15 for the address).

Explanation of Units of Energy and Useful Conversion Information

Calorific Values (higher)

Wood	15 GJ/tonne (approximately) at 20% moisture content
Coal	20–30 GJ/tonne
Natural Gas	44 MJ/m³
Propane Gas	46.3 GJ/tonne
Fuel Oil	45–46 GJ/tonne

Conversion Tables

Area/distance/mass
1 yard (yd) = 0.914 metres (m)
1 kilometre (km) = 1000 m
1 hectare (ha) = 10000 m²
1 km² = 100 ha
1 acre = 0.4047 ha
1 pound (lb) = 0.4536 kilogramme (kg)

Density
1 lb/ft³ = 16.02 kg/m³

Power
1 horsepower (hp) = 0.746 kilowatt (kW)

Timber trade measurements
1 m³ = 27.7361 hoppus feet
1 hoppus foot = 0.036054 m³

Energy
Energy = Power × Time
1 Joule (J) = 1 Watt second (Ws)
1 kJ = 1 kWs
3.6 MJ = 1 kWh
1055 J = 1 Btu
2.685 MJ = 1 hph
1 Therm = 29.037 kWh

1000 J = 1 kilo joule (kj)
1000 kJ = 1 Mega joule (MJ)
1000 MJ = 1 Giga joule (GJ)
1000 GJ = 1 Tera joule (TJ)

1000 Wh = 1 kilowatt hour (kWh)
1000 kWh = 1MWh
1000 MWh = GWh
1000 GWh = TWh

10^3 = kilo
10^6 = Mega
10^9 = Giga
10^{12} = Tera

(continued)

Energy facts and figures (*continued*)

Wood Fuel Production Information

Note: information given is very approximate

Typical arable coppice yield
8–20 dry tonnes/hectare/year, i.e. 24–60 total dry tonnes/hectare on 3 year rotation.

At harvest, moisture content tends to be around 50% and yields in the field of wet wood can vary from 16–40 tonnes/hectare/year.

Bulk density of wood chips
Bulk density of wood chips (tonnes/m³) = 13.6/(100% – % moisture content)
For example: bulk density of wood chips at 50% moisture = 0.272 tonnes/m³

Densities of fresh-felled logs
Willow 705 kg/m³
Corsican Pine 1000 kg/m³
Sweet Chestnut 1040 kg/m³

Densities of seasoned logs (e.g. 25% moisture)
Willow 480 kg/m³
Sweet Chestnut 575 kg/m³

Wood Fuel Energy Conversion Information

1 tonne of wood chips (i.e. 20–70% moisture) can occupy 2–6 m³ (average, depending on chip size, moisture content, loading procedure and settling).

1 tonne wood chips (at 20% moisture) = 15 GJ total energy = 13 GJ useful heat energy.
Very approximately, 1 tonne of wood chips is required for every megawatt hour of electricity (assumes 20% moisture and 25% conversion efficiency).

MEET THE GIGAJOULE

Before we go any further, however, let's discuss the gigajoule — perhaps a word that farmers have not traditionally bandied about. A joule is the internationally agreed unit of energy, equal to the work done when an electrical current of one ampere is passed through a resistance of one ohm for one second. 'Power' is 'flowing energy', or if you prefer it, 'energy rate'. Power is measured in watts. One watt is the power of one joule per second. One kilowatt is 1000 watts and one megawatt is a million watts. The energy value of a fuel is measured in gigajoules per tonne. A gigajoule is one thousand million joules and is written as GJ. As a theoretical example, therefore, if you are burning, say, one tonne of woodchips an hour and it produces 10 gigajoules during that hour, to find out the wattage, you divide the gigajoules by the number of seconds in the hour (3600) and that gives you 2,777,777 watts which is 2.7 megawatts. If you burned the same wood in half an hour, the 'power' would be 5.5 megawatts.

We measure land in acres and hectares, wheat in tonnes and milk in litres. Energy from farm crops is discussed in gigajoules, and power in kilowatts and megawatts. When you are selling energy, talk about £/GJ. (If you still think in terms of British thermal units, Fred Dumbleton of ETSU has a useful tip: 'One million Btu is exactly the same as ten therms and is almost equivalent to one gigajoule,' he says.)

READY MARKETS

If a heat or electricity user in a rural area is not connected to the mains, or is at a distance from a coal supply, home-grown energy may already have a healthy chance of making money. Rural users tend to buy energy in small amounts, which cost more. For example £2–£4 per gigajoule for oil, £6–£7 per gigajoule for LPG and £6 per gigajoule for off-peak electricity.[1] Electricity companies or 'utility companies' which distribute electricity are seeking 15-year contracts with farmers to produce woodchips for bio-fuelled power stations. I understand they have been talking in terms of something like £1.75/GJ. I also understand that, provided you can produce sufficient quantities, and that transport costs between your land and the power station are not too high, then that price can bring in profit equivalent to that of grain crops (at the time of writing). Much will depend on your costs, though.

The greatest expense so far comes from buying the planting material; and, if you are adding value to the crop by converting it into electricity or heat, from building the plant in which to utilise the woodchips. Already the price of cuttings has begun to tumble as suppliers proliferate, with material being imported from Sweden and grown locally (see Chapter 5). At present, the investment needed to start up in coppice production is high and it takes at least two years, probably four, before any significant income is brought in. Admittedly the start-up costs have fallen sharply over the past two years, ever since regular farmers have begun planting willow and poplar. Farmers always find more efficient ways of doing things and short cuts to save expense. In Sweden, I am told, planting material has already come down to the equivalent of 2p or 3p per cutting (from 10p) and establishment costs are down to £500 a hectare already.[2] One should remember, however, that in Sweden they plant twice the number of cuttings per hectare as we do (they are planting up to 20,000 per hectare compared with our 10,000), so their cost per hectare is similar.

If you are borrowing money (and have no grants) to establish coppice, you might expect to pay an 8 per cent interest rate over, say, 20 years. After the first four years, during which time you cannot expect any income from production, here are some estimates as to what your cost per gigajoule might be. I should add that the projected increases in yield will come mostly from the intensive breeding programme being carried out by the Swedes and in the UK. That is one reason for not being 'penny wise and pound foolish' by going for the cheapest planting material: if everyone does that, the breeders will not be able to proceed.

How establishment prices are expected to fall

1995	By year 2000	By year 2015
(yields of 11 dt/ha)	(15 dt/ha/year)	(21 dt/ha/year)
£0.61/GJ	£0.27/GJ	£0.13/GJ

By the year 2015, then, establishment costs should fall to a quarter of what they are today.

CHEAPER HARVESTING AND DELIVERY

To take another example of falling costs, let's look at harvesting and delivery costs. Murray Carter, who farms near Harrogate, gave figures demonstrating the reduction in costs in this area of production when he addressed a conference at the National Agricultural Centre at Stoneleigh organised by the RASE, the Wood Energy Development Group, the NFU and the CLA.[3] He compared harvesting with a special coppice harvester which keeps rods intact, a machine rather resembling a complex maize harvester, developed at Loughry in Northern Ireland, with an adapted Claas forage harvester which cuts and chips in one operation, blowing the chips into a forage trailer.

Cutting with the Loughry harvester cost £7.50 per dry tonne, handling £5.60, chipping £8.35 and transport £5.40, the total coming to £26.85 per dry tonne. Using the Claas, however, cutting and chipping are done both at the same time and the total cost, including transport, was only £11.00 per dry tonne, a reduction of more than 50 per cent in costs. Murray is now talking about an even cheaper way of harvesting in which rods are pulled into a baling chamber (see Chapter 9). He illustrated the importance to the returns of this improvement in harvesting and delivery costs in the following table:

Short-rotation coppice financial performance under two harvesting systems based on a 20 hectare area of which 5 hectares are cut annually

Wood value per dry tonne	Yield/dt/ha/year 1st cut and subsequent cuts	Harvesting cost per dry tonne	Cumulative net profit from 20 ha at year 15
£32	9 and 12	£27 (Loughry)	£12,555
£38	9 and 12	£27 (Loughry)	£37,080
£32	9 and 12	£11 (Claas)	£77,957
£38	9 and 12	£11 (Claas)	£102,483

So, said Murray Carter, when the cumulative net profit figures shown above, from the 15-year short-rotation coppice, were compared with those for cereals the results were as follows:

Arable cumulative net profit from cereals

Net profit per hectare	Cumulative net profit from 20 ha per year at year 15
£150	£79,526
£200	£106,034
£250	£132,543

Thus it is evident that where the Claas forage harvester is used, the financial performance of the two types of crop is broadly similar. Murray Carter went on to say that profitability from short-rotation coppice was likely to rise further as yields increased and as input costs continued to fall. Cost reductions in just one area of production make a large difference to net profits over the long term. John Seed suggests that with average annual yields of 16–22 wet tonnes of woodchip per hectare (that's 8–11 tonnes dry matter) it should be possible to achieve a gross margin of £320/ha. That compares favourably with Scottish Agricultural College figures predicting average gross margins for arable crops or set-aside of £315/ha for 1995/6.[4] John has also been quoted as offering £40 per dry tonne for woodchips.[5]

But let us take a closer look at some figures. Land agent John Lockhart, of Samuel Rose, spoke from his own experience and from that of the five farmers engaged on the Farm Wood Fuel and Energy Project (a project organised for ETSU by Edward Stenhouse and Dr Clare Lukehurst of Clutton & Clutton) across the south of England. John Lockhart has been mainly responsible for the use of woodchips to heat Drayton House, on the Drayton Estate in Northamptonshire, using forest by-products and short-rotation willow coppice. He says that establishment costs are still high, at £1323 a hectare, of which planting material costs £790. However, for farmers who are planting short-rotation coppice on non-set-aside land there is the Forest Authority's Woodland Grant Scheme, offering a one-off payment of £600 per hectare. Better Land Supplement used to be available for short-rotation coppice but this was withdrawn in July 1994. The Woodland Grant Scheme money is only available (at the time of writing) to a maximum of 1000 hectares per year for the five-year period beginning in the summer of 1994. Another source of money to go with the Woodland Grant Scheme is the Community Woodland Supplement, which gives £950 per hectare for the first year only, if there is open public access to the site.

With set-aside, short-rotation arable coppice for energy can be eligible for about £300 per hectare per year for five years (with the £300 a guaranteed minimum for the five years — it could go up). Extra payments of £90 a hectare for access strips and £45 a hectare for whole or part fields could be claimed under the new Countryside Access Scheme[6] and a new grant was introduced in summer 1994 by way of a single payment of £400 after planting. This is restricted to 1250 hectares a year and the scheme will run initially for five years. On the face of it, the limitation of grant to such small areas would seem to inhibit the rapid expansion of home-grown energy from short-rotation coppice. Even one small generating station, such as that planned at Indian Queens in Cornwall, would require more than 2000 hectares of coppice, although the planting would probably be spread over three years.

OTHER WAYS AROUND THE GRANTS

It has been suggested that some planting of short-rotation coppice could be done in such a way as to comply with Forestry Commission parameters for broadleaved tree planting. These grants remain at £1350 per hectare for areas of up to 10 hectares, or £1050 per hectare for more than 10 hectares. This planting must be at the density of 2250 trees per hectare (unless it's for amenity or new native woodlands — which would not apply with prospective coppicing). Therefore, instead of planting a third of your coppice requirement each year, you could plant a proportion of your land straight into poplar, at 2250 trees per hectare, with a view to harvesting it in five or six years. With the market for woodchips increasing it might well pay to plant extra land in this way, to take full benefit of the grants on offer, although I'm told that the cost of harvesting this single stem crop is 'astronomical' because it is 'neither forestry nor short-rotation coppice and doesn't suit harvesting techniques for either'. Another possibility, which would need to be carefully explored with Forestry Commission staff, might be to plant at the rate of 2250 trees per hectare with one forage harvester's width of willow in between.

Do apply for grants *before* starting work! So many farmers have started work in anticipation of receiving a grant, only to be disappointed because the rules are very strict and officials make themselves liable for disciplinary action if they allow exceptions. Contact the Forestry Commission's local representative in the very early planning stages of a project to discover what is allowed, and what is required before spending any money on cultivations, fencing or planting material.

John Lockhart has produced various budgets for establishing short-rotation coppice, based on his experience and using figures from the five demonstration farms in the south of England being advised by Clutton & Clutton. First, establishment costs:

The costs of establishing short-rotation coppice

		£ per hectare
PRE-PLANTING	Cost of herbicide	30.89
	Herbicide application	12.36
CULTIVATIONS	Ploughing	35.83
	Cultivations	27.18
COST OF PLANT CUTTINGS		790.72
PLANTING		239.69
FERTILISER		—
PROTECTION	Post-planting herbicide	81.54
	Herbicide application	23.47
	Additional weeding	44.48
SUNDRIES		37.07
TOTAL per hectare, before grants		1323.23

(He has not included fencing costs in this example — although many plantings will require it for the first year or two.)

Let us compare the benefits of the various grants available. Looking at John Lockhart's budget for the first five years, assuming that non-rotational set-aside has been chosen, the figures are as follows:[7]

ESTABLISHMENT COSTS at £1323.23 per hectare, with money
borrowed and repaid over five years at 9 per cent interest is: £341/ha/year

ANNUAL SET-ASIDE PAYMENT
  Assume £296.52   £296.52/ha/year
NET DEFICIT...£44.48/ha/year

Gross Margin
Annual cost of herbicide treatment after harvest, every
three years at £60/ha   £20/ha

Annual cost of fertiliser: 60 kg of 20:10:10 after three-year harvest   £8/ha

Contract harvesting after three years at £300/ha   £100/ha

Contract haulage at £3.30 per dry tonne within a 10-mile radius   £50/ha

Total variable costs   £178/ha

Output: 15 dry tonnes a year, per hectare at (say) a price of £34. This will produce an income of £510

This would give a gross margin of £332[8]

Total return £287.52/ha/year

Doing a budget for a Woodland Grant Scheme will look less encouraging, however, and the new rates of grant may mean that short-rotation coppice can only be planted on grassland that is too wet for grazing — and is unsuitable for the building of aquaculture ponds or amenity lakes.

John Lockhart tells me that, using woodchips to provide fuel for a stately home, as he has done at Drayton House in Northamptonshire, 'You may not save much on the price of the fuel itself because woodchips may well cost about the same as oil: the difference is that instead of spending money with oil companies, you are spending it with local farmers.' I understand that the house is also kept much warmer for the same cost than when oil was the main fuel!

YES — BUT CAN I MAKE IT PAY?

Whether an individual can make energy from farm cropping pay boils down to one essential: if the electric power distributing companies are offering £1.50–£1.75 per gigajoule, or £30–£38 per oven-dry tonne (which is the range being mentioned in the market place at the time of writing), can energy be produced to meet that price and keep a sufficient margin? We have to start with that price, therefore, and work backwards. The costs of production and delivery will be directly affected by yield, dry matter of the finished product and the cost of inputs, harvesting and transport. John Lockhart says that experience at Drayton suggests that if a market for woodchips can be found outside the energy field you may do rather better, with prices of £15–£20 per cubic metre or £45–£60 per oven-dry tonne being paid for mulch. He believes that with herbicides being restricted on large landscaping contracts and the increasing cost of wood bark, this market is likely to expand.[9] He points out that changes in the price of woodchips make a big difference in gross margins per hectare. A price of £34/dt will bring a gross margin of £332/ha but £38/dt will bring £392/ha and £45/dt will boost the gross margin to £497/ha.

If you are lucky enough to get £50/dt then your gross margin goes up to £750/ha. Price, then, is all important! But what about costs? Establishing the crop can cost £1323 per hectare (but should, in time, cost a lot less). Harvesting costs, as we saw earlier in this chapter, are on their way down but John Lockhart uses a figure of £300 per hectare, including transport to store. Storage and drying are very important costs. If you intend to sell electricity or heat, or both, all year round — and harvesting takes place only between December and February — then storage will be necessary for 9–15 months, if you want to use woodchips with a moisture content of 20–25 per cent. John says that in his experience woodchips are easy to store but it is vital to make sure they are covered to keep the rain off; with natural ventilation woodchips will dry down to 20–25 per cent moisture within four to six months. There is some danger of heating in the heap and it may be necessary to use some kind of artificial ventilation for an initial period (see Chapter 10). John Lockhart considers an opportunity cost of £0.80 per cubic metre of storage space to be realistic. (Banks of Sandy gives a rough estimate for the bulk density of woodchips at 50 per cent moisture of 0.272 tonnes per cubic metre.) 'You can't afford to carry woodchips very far!'

A low density, comparatively low energy value material like woodchips cannot carry much transport cost, so the nearer you can use it, the more likely you are to make it pay. Moved within the village, by tractors and trailers which might otherwise lie idle, woodchips are not going to cost very much at all. But, as John Lockhart says, once lorries have to be hired to take the chips to a generation or heating plant this is going to cost money. He suggests £3.30 per dry tonne or thereabouts, for haulage within a 10-mile radius.

CASE STUDY

John Lockhart takes a case study of a typical 200 hectare arable unit. Looking at the farm without arable coppice but taking 18 per cent set-aside, he calculates a net farm income figure of £84.50 per hectare. The 164 hectares of arable crops should bring in a gross margin of £525/ha, totalling £86,100, while the 36 ha of set-aside at £300/ha brings in £10,800, making a whole farm gross margin of £96,900. With fixed costs of £400/ha, including rent equivalent but excluding finance adding up to £80,000 for the whole farm, the net farm income will be £16,900 or £84.50 per hectare (£33.80/acre). John claims that by planting half of the set-aside to short-rotation coppice on the same unit net farm income could be increased to £104.38 per hectare (an increase of nearly £20). These figures are using a price of £34/dry tonne of woodchips. Put that price up to £38 and the net farm income goes up to £109.78/ha — £25/ha better than the arable crops and set-aside combination.

This is how he works it out. Arable crops take up 164 hectares, bringing in £86,100, and set-aside will yield £10,800 as before. But the arable coppice, on 18 hectares, giving a gross margin of £332/ha will give a total of £5976 if the woodchip price is £34/dt or £7056 if it is £38/dt. That would yield a total farm gross margin of £102,876 (at £34/dt) or £103,956 (at £38/dt). Fixed costs will be a little higher, at £410/ha — because of the extra expense with the coppice — and the total fixed costs will now add up to £82,000, giving a net farm income of £20,876 for the lower chip price and £21,956 for the better price — thus the £104.38 or £109.78/ha claimed.

John Lockhart says that these figures show a useful increase over other non-food crops such as high erucic oilseed rape, which would probably yield a gross margin of £100/ha. He goes on to say, though, that the healthy net farm income is very much dependent on the set-aside money. Without the arable area payments the income would fall even on the higher priced woodchips. Taking 18 hectares for coppice from the 200 would leave 182 ha under arable. Fifteen per cent of this remainder is 27.3 ha, which would bring in £300/ha from set-aside, giving a total of £8190. The 85 per cent under arable crops, which comes to 154.7 ha with a gross margin of £525/ha, yields £81,218. Eighteen hectares of coppice at a gross margin of £392/ha (don't forget it's the higher woodchip price of £38/dt) brings in £7056, giving a total farm gross income of £96,464. Take away the fixed costs at £405/ha, which add up to £81,000, and the net farm income is £15,464 (£77.32/ha). John says that these figures are 'put up to be shot down' and that they are simply a guide for farmers and landowners who can substitute their own costings.

With the new grant of £400 per hectare in the first year of set-aside, the figures will look healthier — if you manage to get your share of the 1250 lucky hectares given grants each year.

'Never lose sight of the aim — which is to come up with the lowest cost per dry tonne delivered: and this is not necessarily equal to the largest yield per hectare,' said Edward Willmott, the director of WoodGen Ltd, speaking at the autumn conference 'Coppice — looking beyond set-aside', held at the National Agricultural Centre at Stoneleigh in October 1994.[10] He was referring to factors which increase the cost of woodchips, such as using land with poor harvesting access or where the machines get bogged down, or long journeys to the point of processing — even though yields at the field are very high.

The cost that dominates the economics of short-rotation coppice is that of establishment, Willmott told delegates. It could be as high as 30–50 per cent of the cost of the fuel produced. He gave the following example from one of the projects entered for NFFO 3, but said the calculations could equally apply to crops of the future, like miscanthus.

To establish one hectare of short-rotation coppiced willow

Cost of cultivations	£82.50
Planting and labour	£250.00
Sprays etc.	£113.50
Fencing etc.	£200.00
TOTAL	£646.00

Then add cuttings, at 10,000 per hectare, first at 9p each. That costs £900 per hectare, bringing the total establishment cost to £1546 per hectare. However, using cheaper cuttings — say at 4p each, also planted at 10,000 per hectare — produces a total cutting cost of only £400 per hectare and brings the total establishment cost down to £1046 per hectare — a reduction of nearly a third! The choice would be up to the farmer. Personally he worried that new, expensive clones might not live up to expectations and, if given the choice, he would follow a policy close to his heart of 'not spending money on day one'.

COSTS OF ADDING VALUE

The cost of boilers (and gasifiers) is beginning to fall too, as more companies realise the increasing importance of home-grown energy in national and international policy. This is not to deny that costs for fuel from short-rotation coppice are still very expensive compared with coal, gas or oil boilers. Every company seems to have a gasifier (*the* answer to woodchip utilisation, they say) 'coming on to the prototype stage', and some claim to be 'there already'. Farmers will judge whether gasifiers are ready to provide a completely reliable service — they should not be touched with a barge pole otherwise, because of penalty clauses built into heat and power supply projects. Hoppers, burners and boilers suitable for woodchips seem to be over-priced at present, probably because, until now, fuel has been very cheap or even free. Woodworking factories have had problems disposing of sawdust and woodchips — and so could afford to pay premium prices for equipment which would convert an embarrassment into useful heat. With wood-chips increasing in price, processing plant must fall in price if it is to sell.

WAITING FOR HARVEST

If you are going to add value to the woodchips produced on your own land, you face another cost which could prove very difficult to cope with in the first three or four years. If you draw up a contract to sell heat, or heat and power to some institution, to a group of houses or an industrial site, your customer will want the product as soon as possible. They are not likely to want to wait until the short-rotation coppice has grown. You, therefore, have to find an instant source of woodchips. I claimed in Chapter 1 that the market for woodchips is widening and that this material is required for more than just energy. This means that woodchips may well be fetching quite high prices, unless you are near a timber mill or joinery factory where disposal of sawdust and woodchips is an embarrassment to the operation. It is likely that you may have to do as Rupert Burr is doing, near Swindon — coming to an agreement with local woodland owners to tidy up their woodland in order to acquire wood-chips. This will not be cheap, for the woodland owner will want his or her woodland left neat and tidy, without great ruts and sloughs where heavy machinery has been used at the wrong time of year. Also actual chipping needs to be done. Tractor-mounted chippers do exist, and work well for small quantities and odd jobs, but something larger is needed if you are looking to supply, say, 300 or even 1000 tonnes of woodchips a year to produce the power for your customer.

There are forestry harvesting contractors, who can be con-tacted through the FCA (Forestry Contractors Association); the address and phone number will be found in Chapter 15. There is also at least one 'visiting chipping service' available. Gannon UK Ltd runs a 'milk round', rather in the same way as the old threshing machines used to do a regular circuit in the winter. Gannon charges £800 per eight-hour working day, within a two-hour transport radius of its base in Lincolnshire. The com-pany runs very heavy duty industrial grinders which have a 10-foot diameter tub and can easily take 8-foot cordwood. Out-put is up to 15 tonnes per hour if the wood is properly prepared and fed by a grapple or telescopic loader.

UNFAIR COMPETITION

There is a political dimension to the cost/benefit analysis of home-grown energy which cannot be ignored. It will be up to potential producers to pursue profitability through political and parliamentary channels. At present, the production of heat and power from home-grown energy is having to compete against cheap heat and power which is being sold regardless (almost) of the consequences to the future of the world. Cheap power now, extinction later — heat and light now, for tomorrow we die! Nuclear power *is* fantastic — but how happy are you that nuclear waste should be stored under your house or farm, not just for your lifetime but for that of your children, grandchildren and great-grandchildren? Coal, oil and natural gas *are* so easy, so controllable — but are you prepared to pay (and pay hand-somely) for all that carbon dioxide to be converted into some-thing solid and useful rather than being tipped into the atmos-phere, ultimately changing the climate so that millions can no longer make a living? Are you prepared to pay for all the sulphur and other damaging materials to be removed from exhaust gases? Not until you have to, forced by the law. Once again we are back to political decisions.

British politicians have set out their view of the future in 'New and renewable energy: future prospects in the UK'. Published in March 1994, it is very bullish about the economic prospects for energy crops. It assumes that yields from coppice will rise to 21 tonnes of dry matter per hectare per year by 2010; and as gasification comes of age, efficiency of conversion will increase from 1992's level of 25 per cent to 35 per cent, also by 2010. The DTI's paper says that projected increases in coppice yield and power station efficiency coupled with reductions in establishment and harvesting costs will lead to electricity production costs comparable to new, coal-fuelled power stations. The prospects for bio-ethanol or bio-diesel are not so encouraging: the paper says they are likely to become competitive only in niche markets or with heavy subsidies.

Government intends to reduce establishment costs to farmers and landowners through set-aside grants and Forestry Authority woodland grants. This, together with improved planting regimes, should cut through the restraints on technology which are daunting prospective growers. Government's intent is plain: 'To encourage the uptake of energy crops', using a range of incentives and providing research and technical back-up. One thing is sure, governments are no longer prepared to offer open-ended subsidy committments to farmers, although the DTI paper indicates clearly that energy crops will receive various forms of pump-priming, notably through the NFFO and probably by more establishment grants.[11]

THE BOTTOM LINE

The competition facing energy crops is at present completely inequitable because inherited resources (oil, coal and natural gas) are continuing to be squandered at an unprecedented rate, regardless of the consequences, which include global damage. Until other forms of energy are properly paid for, including charges for recycling their by-products, making money from home-grown energy will always be a struggle. Immediately proper charges are made for existing forms of energy, however, home-grown energy becomes economically viable.

Take Sweden, for example, where a tax on fossil fuels has instantly put short-rotation coppice into the black: many thousands of hectares of coppice are now in full production and power stations and district heating plants are running profitably. Land is being put to good use; fossil reserves are being conserved; and the environment is being protected.

Finally, there is some discouraging news on the home front. I am told, on good authority, that in the landfill gas programme, which is a parallel field to biomass production for fuel, once a landfill operation is extracting gas to displace other fuels, especially natural gas, the price of the fossil fuel being sold to current users mysteriously and suddenly drops so much that the landfill gas becomes uneconomic. This kind of unfair competition could very well be brought into play when wood fuel producers potentially oust a number of other players in the existing energy market.

References

1 ETSU, 'An assessment of renewable energy for the UK', HMSO, 1994.
2 Personal communication with John Seed, 27 April 1994.
3 Murray Carter, 'The future: keys to progress', proceedings from 'Short-rotation coppice — growing for profit' conference, Wednesday 24 March 1993 at the NAC. Published by RASE and WEDG.
4 John Seed, 'Fuel for the future' (leaflet), 1993.
5 J. Williams 'Coppice fuel developments', *Farm Development Review*, 1994.
6 S. Gregson, 'Increased support fuels progress on energy crops,' *Farm and Country* magazine, RASE, March 1994.
7 John Lockhart, paper given at Banks of Sandy farmer meetings at Long Melford and Kettering, 1994.
8 See note 6.
9 Personal communication, 6 May 1994, Calne.
10 Edward Willmott, 'Theory into practice', paper at RASE, WEDG and BB conference, supported by the CLA and NFU, 'Coppice — looking beyond set-aside', 5 October 1994.
11 DTI, 'New and renewable energy: future prospects in the UK', Energy Paper No. 62, 1994.

CHAPTER 5

Planting Material

Spend as much as you can afford on planting material, because short-rotation coppice is a crop which could stay in the ground for up to 30 years, as we shall see later in this chapter. The nearest farming equivalent is establishing a pedigree herd of cattle. You would not start with cows bought at random from the local cattle market. You would, instead, do a lot of homework to discover upon which herds you will base your own.

Planting material arrives from the cool store ready for planting and painted at the top to show which way is 'up'.

The plant material to buy is unrooted cuttings from 'maiden' (one-year-old) stems. With willow, cuttings are made by cutting shoots to the required length of 20–25 cm, but for poplar the conventional forestry practice has been to use only unbranched material and to cut each piece to an undeveloped primary bud. For short-rotation coppice, however, you need so many cuttings that such a technique would be wasteful and time-consuming, and recent experience has shown that cuttings taken from shoots from which side branches have been trimmed (and hence lacking primary buds) are markedly slower to develop or may even fail to grow at all. Research work is going on to see how serious this is and to seek solutions, such as possibly leaving sufficient side branch untrimmed to include undeveloped secondary buds. Short-rotation coppice specialist Malcolm Dawson, at DAFS in Northern Ireland, however, maintains that he would stick to primary buds or unbranched maiden stems. 'Any stepping back from this reduces quality,' he says.

Willow and poplar are both members of the *Salicaceae* family. (Just to remind you, life is divided into kingdoms, phyla, classes, orders, families, genera and species.) The genus for willow is *Salix* and is divided into three sub-genera, the first being *Salix*, such as the type from which we get cricket bat willow, which is *Salix alba caerulea*. Then there is the sub-genus *Salix chamaetia*, representing the very small willows such as arctic and creeping willows, not suitable for coppicing; and finally there's the sub-genus *S. caprisalix vetrix*, the bush or shrub willows. This sub-genus contains broad-leaved willows such as *Salix caprea* (the goat willow), *Salix cinerea* and the narrow-leaved *Salix viminalis*, or basket willow, which grows straight, mainly upright rods such as we need for coppicing. If you wish to get to know the botany — and much more — of willows consult *Willows – The Genus Salix* by Christopher Newsholme, recently published by Batsford.

EARLY DAYS

Although selection programmes for willow have been going on for 15 years in this country, it is still very early days, according to Malcolm Dawson of the Northern Ireland Department of Agriculture, one of the pioneers in this field. While he agrees that we have suitable types of willow to get the industry off the ground, he insists that if we wish to create a large-scale energy programme based on willow and poplar an ongoing breeding programme is needed to develop clones suited to the UK's maritime climate.

'Where you are planting whole stands of just one type of clone, the life of that clone will only be about ten years, in our experience,' he told me. In Northern Ireland what has happened is that, after that kind of period, the clone has gone down to the rust disease *Melampsora*.

'You *must* plant polyclonal stands!' Malcolm Dawson says (by which he means mixtures of clones). He suggests a minimum of five different clones in a mixture but 'the more the merrier'. 'It will not only increase your yield, as the clones intercept the available light entering the crop more effectively; but there is also a dilution effect and the selection pressure the system exerts on the disease organism is reduced. Diversity creates stability.'

There is less opportunity for a disease to build up. It's not yet clear which is the best way to mix plants — whether to have one row of one clone, and then one row of another, or whether to have random mixtures over the whole field. At Long Ashton and Loughgall they are experimenting with both row-by-row and random mixes. Until now they have only been using up to 6 clones in a mixture, but recent plantings contain plots with 5, 10, 15 and 20 clone mixes. 'It is not, however, just any old mix,' says Rod Parfitt. 'Clones must be chosen to be not mutually antagonistic (e.g. if one is later starting into growth it may be shaded out); and also, more importantly, not to be mutually susceptible to some rust races. A wrong mixture may exacerbate problems.'

Just what constitutes a 'right' mix is the subject of research being conducted by Dr David Royle at Long Ashton. Malcolm Dawson says the industry needs a rolling programme of selection and breeding, to keep a stream of new clones coming on to the market, so that disease can be kept at bay. At Loughgall, where he works, they have been running trials of new varieties since 1990 and about half the new clones being examined are rejected. This service will prove very necessary to large-scale growers of willow throughout the UK, and it could be carried out by individual breeders as well as by government stations or ADAS.

VARIETIES ON THE MARKET

Malcolm Dawson says there are plenty of good, old varieties of willow on the market. All varieties carry the risk of breakdown to rust, but this risk is less on new material because it hasn't been around as long. The very latest planting material has been selected for what Malcolm calls 'field resistance', or partial tolerance to rust. It may not be *free* from rust but it can live with it, without too much loss in yield. This type of field resistance is potentially more sustainable than total resistance.

The UK 'standard' variety is *Salix viminalis* 'Bowles hybrid'. This is decades old and was a product of the basketmaking industry. It was almost certainly produced by the eponymous Mr Bowles, who was still alive at the turn of this century. It has good yield, straight, upright rods and is persistent. It can suffer from rust, but in an unusual way; normally rust affects only the leaves of willow, but with 'Bowles hybrid' it affects the stem too, leaving pustules there. Nevertheless, says Malcolm Dawson, it may be a good idea to include 'Bowles hybrid' in mixtures in the areas where disease pressure is not so strong. In Northern Ireland, however, the cooler damper weather encourages rust.

Interestingly 'Bowles hybrid' may be the source of a biological control for rust; work is going on at Long Ashton Research Station to pursue the matter. *Sphaerellopsis filum*, a fungal hyperparasite of *Melampsora* rust, overwinters in the stem cankers found on 'Bowles hybrid', providing a reservoir of inoculum to infect leaf pustules early in the season: there is a chance that it may be possible to breed up this fungal disease to counterbalance rust.

Another old favourite is a hybrid of *Salix viminalis, S. caprea* and *S. cinerea* called *Salix* × *dasyclados* (× denotes a cross or hybrid). 'This gives reasonable yields on many types of site,' says Malcolm Dawson. 'Its parents are native types here and it's a rough and tough clone suitable for north or south.' This hybrid has the disadvantage that it is rather untidy in its growth habits: it develops shoots which grow out from the stool and then bend upwards (like a hockey stick) instead of growing up straight, in military fashion. This may become less important as new harvesting methods are developed. Unruly growth used to be a 'deadly sin' for willow, when rod harvesters alone were used, because it meant some side shoots avoided the harvester, leaving a messy field. But with forage harvesting techniques or with the massive Swedish coppice harvesters now available, side shoots get gobbled up with the rest.

Such shoots, however, will always be something of a problem at harvest. There's one very useful clone whose name is disputed by the taxonomists: they can't agree about what it ought to be called, but the general name given is *Salix burjatica* 'Germany'. There are other varieties, including one called *S. burjatica* 'Korso' and another called '81090', but Malcolm Dawson does not recommend these because of their susceptibility to rust. *Salix burjatica* 'Germany', however, although it gets rust, gets it later in the season and this does not affect yield too seriously. It is an extremely heavy yielder and Malcolm recommends its inclusion in a mixture. On trial plots it has produced 16 dry tonnes per hectare per year, although this, of course, cannot be guaranteed on a field scale.

Then there are several *Salix viminalis* × *triandra* hybrids, notably 'SQ 83'. Developed in Sweden, it has the highest yield of the old planting material, with plot yields of up to 18 dry tonnes per hectare per year — but again it has the disadvantage of being rather untidy, especially on outside rows. Another Swedish clone, a *Salix viminalis* called '78183' comes from the University at Uppsala. (These numbered names are based on the year when crosses were made, followed by the number of the plot: in this case 1978, plot number 183.) Whereas 'Bowles hybrid' is our standard variety in the UK, '78183' is the standard for Sweden: that's the one that other clones are compared with. More than half of Sweden's short-rotation coppice consists of this clone, but it has not adapted well to the UK, and has yielded poorly in trials. The Swedes do not need to go in for mixtures yet, says Malcolm Dawson. Their climate is not conducive to the survival of rust: the winter cold kills it off, and in summer temperatures sometimes rise to 25 degrees Celsius, which inhibits spore development.

Malcolm Dawson does not recommend *Salix* × *hirtei*. There is *Salix* × *hirtei* 'Delamere', 'Reifenweide' and 'Rosewarne White', but all are susceptible to rust. He prefers *Salix spaethii*, of unknown origin, which is nice and straight, has shorter, fat stems and although it does get rust, is partly tolerant to it. *Salix viminalis* × *Salix triandra* '2481/55' is a relative of 'SQ 83' and you can use it in the same mixture as 'SQ 83' because it has a different genetic make-up. *Salix × stipularis* is one of the old varieties, which keeps surviving: it is straight and very tall, and Malcolm would include it in a mixture.

There's a list of *Salix viminalis* clones, like 'Bowles hybrid', 'Campbell 3106', 'Gigantea', 'Q 683', 'Lysta Q 699' and 'Mullatin'. Malcolm Dawson recommends all these. Another he would include is *Salix × calodendron*, which has the same parents as *Salix dasyclados* but different genetic material, with different parents predominant. There are *Salix caprea × Salix viminalis* hybrids and two of them are very, very straight, have broader than usual leaves, are not as tall as *S. viminalis* but have thicker stems. They are *Salix × sericans* 'Niginians Prunifolia' and *Salix × sericans* 'Coles'.

Those, then, are the basic varieties which should now be widely available in the UK, from various suppliers. It would pay to check out your supplier before ordering. But there is also newer material which has been trialled in this country. Malcolm Dawson gave me a run-down on this first generation of systematically organised breeding varieties, developed in Sweden especially for short-rotation coppice, which should contain 'the beginnings of resistance to rust disease'. They are all *Salix viminalis* clones: '78112'; '78118'; '78021'; '78195'; '78176'; '85002', also known as 'Gustav'. The next generation, '870082', '870083' and '870148', again all *Salix viminalis*, are the 'Rapp', 'Orm' and 'Ulv' of the Swedish connection.

In the coming years the rush to get hold of planting material may mean that growers might have to buy these clones direct from Sweden; note that these particular ones have been trialled in the UK. Other clones which have proved very useful in Sweden have been rejected for use here: buyers should make sure that what they are planting will be suitable for their purposes and that their supplier will provide the necessary advice and services. Breeding and selection continues: some types of willow have attributes which breeders would like to include for western Europe. For example, one willow from Siberia, called *Salix schwerinii*, has fewer but thicker stems which may suit certain conditions here, but it has the disadvantage of leafing up in January, laying itself open to frost damage. Types have to be selected for timely closedown in the autumn, so that they won't be affected by early frosts. The willow needs to drop its leaves in good time and harden up, ready for winter.

During the next few years yields are set to increase. Rod Parfitt tells me that in the winter of 1993/4 he harvested some two-year-old stems from stools which had been planted in 1991, cut back in 1991/2 and grown on for just two years. From 10 random blocks of 10 stool plots, he got calculated yields of around 18 oven-dry tonnes per hectare per year (i.e. 36 oven-dry tonnes per hectare at harvest). 'The stools have resprouted magnificently,' he says, 'and by June 1994 were over four feet high!' Rod does not pretend these are commercially achievable yields, but considers that 15 oven-dry tonnes per hectare per year, and perhaps even more, should soon be achievable.

Lionel Hill disappearing into an artificial bank of plastic mesh, soil and mixed willow clones — the perfect sound deadener for motorway noise.

One producer of planting material who has found a good use for some of those 'unruly' types of small willow with a creeping habit is Lionel Hill of Redditch in Worcestershire. He has found a market for them: planting on artificial banks erected to deaden the sound from motorways. The willows form very dense growth and prove most effective. He also supplies parks

and gardens, colleges and schools with willow for 'art' purposes, basketmaking and for growing live mazes. People who
weave growing willow into tunnels, igloos, summerhouses and
other most attractive designs go to Lionel for the appropriate
type of willow and poplar cuttings. Ken Stott, formerly a willow
specialist from Long Ashton, is continuing some of his willow
breeding programme on Lionel Hill's farm. Lionel, along with
Edgar Watson of Claverham, Avon, also supplied most of the
planting material for the five demonstration farms across the
south of England mentioned in Chapter 4.

Lionel Hill has some 60 different willow clones and 30 poplar
clones and says he can supply suitable mixtures of clones for
commercial production. (On his farm he also has a trial willow-
powered, automatically stoked heating system for his house and
indoor swimming pool. He is working with Bob Talbott, of
Talbott Heat Ltd, on this project.)

THE SWEDISH CONNECTION

Yorkshire willow grower and cutting producer Murray Carter is
closely involved in two major breeding programmes. One is
about to start with Long Ashton Research Station, and the
other is already being run in Sweden by the seed company
Svalöf Weibull. The Swedish programme includes some selections from Murray's own farm near Harrogate, sent to Sweden
in the late 1980s. As well as these, there are selections from all
over western Europe and across Russia, right through to the
Pacific coast. Svalöf Weibull, for example, gathered clones of
willow in the Kirov region of Russia, in 1989. In 1990 they made
collections from Novosibirsk in Siberia to the river Amur, close
to the Chinese border. Some 700 clones have been amassed
altogether and crosses are being made between the best clones
of both the old and new collections.

Seven new clones were due to go onto the market in 1994/6,
from the Swedish breeding programme. They were extensively
trialled by Murray Carter in the UK and are called 'Orm', 'Rapp',
'Ulv', 'Jorr', 'Jorunn', 'Tora' and 'Björn'. ('Rapp', the willow, is
not to be confused with 'Rap', which is a clone of poplar.) 'Rapp'
has not done as well as the other two in this country, as you can
see in Tables 5.1 and 5.2, taken from Murray Carter's paper in

the proceedings of the autumn conference 'Coppice — looking beyond set-aside' held in October 1994 at the National Agricultural Centre, Stoneleigh. Murray commented that the results of trials in the UK highlighted the importance of testing new introductions before commercial release here. For example, the reference clone '78183', despite being popular in Sweden, is greatly outyielded by 'Bowles hybrid', 'Orm' and 'Ulv' at Long Ashton and by 'Ulv' and 'Orm' at Loughgall. Similarly, 'Rapp', which frequently outperforms 'Orm' and 'Ulv' in Sweden appears less productive under UK conditions, while 'Ulv' yields are higher. 'Jorr', 'Jorunn', 'Tora' and 'Björn' got off to a very good start in the UK. Look at Table 5.3 to see how quickly they grew. Svalöf Weibull claims that these seven have improved resistance against rust and insects. They are protected by international plant breeders' rights.

Table 5.1   Long Ashton Research Station

Yield expressed in oven-dry tonnes per hectare per annum		
Clone	odt/ha/pa	Yield as % 78183
Bowles hybrid	18.39	146
870082 Orm	18.25	144
870148 Ulv	17.62	140
870083 Rapp	14.09	112
78183	12.63	100
LSD (p = 0.05) = 1.90		

Table 5.2   Department of Agriculture, Northern Ireland, Loughgall

Yield expressed as mean stool dry weight in kilograms		
Clone	Mean stool dry weight	Yield as % of 78183
870148 Ulv	1.74	127
870082 Orm	1.55	113
870083 Rapp	1.46	107
Bowles hybrid	1.37	100
78183	1.37	100
LSD (p = 0.05) = 0.18		

Table 5.3   Comparison of clonal growth rates

Mean shoot lengths as % of 78183

Clone	Planted 1994 One season's growth	Planted 1993 One season's growth on two-year-old stools	Average increase in shoot length
78183	100% = 1.418 metres	100% = 1.850 metres	+0%
Bowles hybrid	126%	125%	+25.5%
Orm	134%	111%	+22.5%
Ulv	140%	101%	+20.5%
Jorr	145%	142%	+43.5%
Jorunn	147%	151%	+49%
Tora	167%	163%	+65%
Björn	170%	171%	+70.5%
	Exmoor trial site	*Harrogate trial site*	

Mean shoot lengths of Bowles hybrid, Orm and Ulv are broadly comparable, showing a 20–26 per cent increase over 78183. Jorr and Jorunn are significantly more vigorous at plus 43–49 per cent and Tora and Björn provide an outstanding increase of 65–71 per cent in one year's shoot growth, although they have fewer stems per stool than pure *Salix viminalis* clones.[1]

Svalöf Weibull, working with the willow specialists at Long Ashton and Murray Carter in Harrogate, will be using planting material from many countries in a programme of structured breeding, in which new varieties are built up from genetic combinations into new, high performance types of willow. Murray Carter is Svalöf Weibull's partner in this country and he tells me that in their breeding programme some 15,000 new clones are produced every year. The best of these — perhaps around ten a year — are brought to the UK for trialling. He has a network of 12 trial sites throughout the UK, from the Orkneys in the north to south-west England, and from the west of Northern Ireland to East Anglia. These sites are on research stations and farms. Plans for another network of trials are being considered by ETSU, with the participation of many in the new home-grown energy industry.

There are about 300 species of *Salix* (willow), of which 30 grow naturally in Sweden. Svalöf Weibull began its willow breeding programme in 1987 to increase biomass yield and improve resistance to insects, fungal diseases and frost. It has trial sites in Sweden, Denmark, Northern Ireland and Great

Britain. Murray Carter has some 60 hectares of trials and propagation beds, with 100 willow and 20 poplar clones under evaluation in collaboration with seven different research organisations. One of these is, of course, Long Ashton, and at this station near Bristol Rod Parfitt runs the trials, working until recently with John Porter who has moved to take a chair in a Danish university.

Although 'Bowles hybrid' still compares very well, in terms of fresh yield, with the new Swedish hybrids, Rod says that moisture content varies greatly between clones. 'Bowles hybrid' may come in at 57 per cent moisture from the field, whereas 'Orm', 'Rapp' and 'Ulv' may have only 52 per cent moisture. 'No longer,' says Rod, 'can breeders and people trialling willow just harvest the crop and apply a "dry matter factor": they must also do the oven-drying tests.' These, he says, are very expensive indeed because there is so much volume of material to be dried. Although the best 25 or so old clones that we have in Britain might appear to pale beside the new Swedish hybrids, they may still be needed for a while, according to Rod Parfitt. 'As we go into the planting of clone mixtures they will have an important part to play,' he says. 'We have to use the *right* mixture for a situation, using clones which have a suitable spread of resistance and tolerance to disease — and tolerance to each other. These mixtures might well need both old and new clones, so that maximum utilisation of light, for example, is made; and some of the old clones may be able to form a partial barrier to a particular type of rust.'

Rod Parfitt says that the new hybrids are difficult for the uninitiated to distinguish by eye, except perhaps 'Orm' (which means snake in Swedish), which has a very slightly wavy stem, and 'Björn', which has reddish tips, but also subtle variations in bark colour between green and reddish brown, and in the colour and shape of the leaves. What makes new hybrids stand out is their vigour, upright growth, lack of side branches and their compact bases. 'They are designed for mechanical harvesting,' says Rod. There are hybrids between superior *Salix viminalis* clones and a species which has not previously been included — but he's not telling which one that is, at present. The new clones should be much more tolerant of disease, according to Rod Parfitt: 'Resistance can break down but tolerance is less likely to.'

POPULAR POPLARS

One of *the* people to talk to about short-rotation coppicing of poplar is Paul Tabbush of the Forestry Commission. He is based at Alice Holt, near Farnham in Surrey, and is quite passionate about poplar, although not quite as enthusiastic about coppicing. He considers the cutting of poplar trees before they have reached their prime rather like killing pedigree calves for veal, but is full of ideas for variations on the theme of coppicing. 'Every time you cut a tree down to its base you are giving it a severe setback,' he says. 'You are also letting light hit bare ground — wasting solar energy.' But he also believes that coppiced trees may well have a very long life indeed.

'I don't know where they got this idea of a useful life of 30 years,' he says, 'because at Westonbirt Arboretum in Gloucestershire we have a lime tree stool which is now many metres across, which has been coppiced for the past 2000 years!' For centuries, he says, fibre from lime bark was used to make rope and the round timber was used for building poles and for firewood. Speaking as a forester, he would prefer to plant at wider spacing and leave the rotation longer. This proves much cheaper at planting time, with far less to spend on cuttings, although it could be said to lead to much more expense at harvest time. This method may not bring in the cash as soon, he says, but will produce a much higher yield and what is harvested has a much wider choice of market, as woodfibre, pulpwood, logs — and, of course, woodchips for the energy market. I am told that the answer to Paul Tabbush's question about the useful life of 30 years came from withy beds on the Somerset Levels where, apparently, it was found, that performance dropped off after that time and that some stools died. Of course today much better varieties are becoming available.

Paul Tabbush thinks people considering short-rotation coppice should look at the possibility of planting part of their ground very densely, with up to 20,000 plants per hectare, to bring in some early cash, say in year two or three, for woodchips. Then another area could be planted at 10,000 cuttings per hectare, another at 2500 per hectare and yet another at 1100 per hectare. This, he believes, could give many more harvesting options, if anything were to go wrong with the woodchip market. 'Many domestic fuelwood markets will demand small logs rather than chips, so it will make sense to plant wider and wait a couple of years longer.' Bearing in mind that woodland planting grants are now scarce for short-rotation coppice, but are still available in full for wider spaced poplar plantations, the extra wait may indeed be worthwhile.

POPLAR ORIGINS

Trees of the poplar genus occur all around the northern hemisphere. On the western side of north America, up to Alaska, *Populus trichocarpa* is found, and is easily recognisable because of its dark green leaves which are white underneath and are longer than they are broad. *Populus trichocarpa* produces somewhat weaker timber than classic black poplar (*Populus nigra*) hybrids, such as have been grown extensively in Europe since the seventeenth century, says Paul Tabbush. The main type of poplar to the east of the Cascade Mountains and in the south of the United States is *Populus deltoides*. 'This is better adapted to drier conditions and warmer weather,' says Paul.

'In Europe we have *Populus nigra* and here in Britain there's a sub-species called *Populus nigra betulifolia*, the now rare "black", or "water" poplar which is stimulating so much conservation activity at present.' It was an inhabitant of native flood-plain woodlands, a habitat type now almost completely lost in Britain.

'In central Asia there's *Populus simonii*, and in Japan *Populus maximowiczii*, and these are all closely related and will interbreed. This means you can get hybrid vigour by crossing them,' says Paul Tabbush. The hybrids used for short-rotation coppice are such crosses. As could be expected, the *Populus trichocarpa* crosses (with American west coast parentage) are better suited to wetter areas, such as Ireland and the west coast of Britain. They are very vigorous in higher rainfall areas although, like all poplar, they don't like too much exposure to violent winds and weather.

'*Trichocarpa* crossed with *deltoides* brings you plants with extra vigour and very large leaves,' says Paul, but he emphasises that there's still much breeding to be done and even better crosses may be found. Poplar will tolerate most soil types, even those with high pH, as long as the soil is deep and well drained, but with a high water-holding capacity. It will not tolerate water-logging. 'Flooding, yes, but not waterlogging.' If a choice has to be made between willow and poplar, the final decision ought to come down to economics and yield.

Modern breeding programmes for poplar have been going on for much longer than those for willow and, as with wil-low, there is huge genetic variation available for crossing. The new Belgian clones are highly resistant to disease. With poplar breeding having gone on for 40 years, Paul Tabbush believes that it has a distinct lead over willow but 'if it's *very* wet, then go for willow — although if it's flat alluvial land near a river, then poplar will do as well, if not better than willow, yieldwise.' He gives the example of one of the five short-rotation cop-pice demonstration farms, that of Robert Goodwin at Kelvedon in Essex: 'The Beaupré poplar there is doing better than the willow.'

Yorkshire coppice-cutting producer Murray Carter disputes much of this. He remarks that the 40 years of poplar breeding were for timber production rather than coppice — and he disagrees about planting willow in preference to poplar in very wet conditions. His advice is to 'check trial results — see what the facts are'. So, once again, when deciding what to plant, grow comparative plots of willow and poplar types, to discover what happens and which trees and types suit your land and condi-tions.

Unfortunately, as with willow, there are no types of poplar that favour shallow soils on chalkland. If you are going to set up trial plots of poplar there are, at the time of writing, some nine types which might be included, says Paul Tabbush. 'For any given site there are probably five that should be tried.'

First there are the two Belgian-bred sisters 'Boelare' (pro-nounced Boo-lar) and 'Beaupré'. They have a *Populus trichocarpa* mother and a *Populus deltoides* father. These trees do not like being exposed to strong winds. They are very vigorous hybrids with extra large leaves ('they're the size of A4 paper,' says Paul) at mid-season. They grow in most parts of the UK and are very resistant to canker and rust. They are not particularly frost susceptible although in a bad year damage can be caused by early autumn or late spring frosts . This is not often a problem, though, and very successful stands of poplar can be found in

Yorkshire and Scotland. 'Most dangerous is a warm March followed by late frosts.' 'Boelare' favours the south of England.

'Beaupré' is very similar to 'Boelare' but there are genetic differences which help to dilute the disease risk. 'Beaupré' is perhaps a little more resistant even than 'Boelare', according to Paul Tabbush. 'But both produce magnificent timber trees if they are allowed to grow on,' he says, harking back to his forester's instinct to let a tree achieve its ultimate glory. He gives an example of 'Beaupré' in a research trial at Bedgebury pinetum in Kent: 'It was planted as a cutting in 1987 and is now 14 metres tall!' (He tells you this with pride, rather as a motor enthusiast gives speed figures of 0–60 in six seconds.)

THE THREE 'Gs' AND THE 'Ts'

For warmer and drier areas there are three candidate poplar types. They are 'Ghoy', 'Gaver' and 'Gibecq'. All are hybrids of *Populus deltoides* and *P. nigra*. They have smaller leaves and are more tolerant of wind, although still not liking over-exposure. They grow slightly more slowly than the 'B' sisters. If you live in the south or east of England these should be included in your trial plots. They have high disease resistance. Then there are the pure *Populus trichocarpa* clones including 'Fritzi Pauley' (the mother tree of 'Beaupré' and 'Boelare'), a female line which has been around for a very long time, according to Paul Tabbush. Not only that, but it's still one of the best. Along with 'Trichobel', 'Scott Pauley' and 'Columbia River' it is highly resistant to disease and will tolerate comparatively acid soil conditions. All these like high rainfall and may be good choices for short-rotation coppice in the western areas of the UK.

' "Primo" doesn't look too promising so far,' says Paul, 'because it just hasn't done well enough in trials and is not sufficiently vigorous.' This, he emphasises, is just his personal opinion, because all these clones are fairly new and there are many more exciting poplars in the pipeline. Paul Tabbush does not favour mixed stands of poplar at this stage. He prefers what he calls 'mosaic' planting. 'You cannot predict what is going to happen with an intimate mix,' he says. 'If disease did get in, and a whole lot of stools died, you wouldn't easily know which type had fallen to the disease.' Malcolm Dawson, in Northern

Ireland, confirms what Paul Tabbush has to say about poplar: 'Poplar trees like sunshine and light. They do not like being exposed to high winds and bad weather: therefore the further south in the UK you can grow them, the better they like it,' he says.

The climate in Northern Ireland is not ideal for poplar but Malcolm does have a few poplars in trials. The two that look best are 'Boelare' and 'Beaupré'. He is pleased if he can get poplar yields of 8 dry tonnes per hectare per year in the first three-year rotation of such hybrids, but maintains that much higher yields could be achieved further south. Some of the poplars that have done best in Northern Ireland are the old *Populus trichocarpa* types, with 'Fritzi Pauley' in particular performing well, despite the moist climate which is reminiscent of the west coast of North America.

European legislation has led to the 'Forest reproductive material regulations, 1986' being adopted by Parliament in the UK. Planting material from some named species of trees, including poplar but not yet willow, may only be marketed if it comes from registered stock. Poplar is included because it is potentially such an important source of good timber. To be able to sell cuttings of poplar you must have the stool beds registered by the Forestry Authority. In 1990 the Forestry Commission Research Division announced the approval of six improved poplar clones bred in Belgium at the Poplar Research Centre at Geraardsbergen by Vic Steenackers; they are: 'Primo', 'Ghoy', 'Gaver' and 'Gibecq', along with 'Beaupré' and 'Boelare'. In 1991 another two clones, 'Trichobel' and 'Columbia River', were added to the list. They were also from Geraardsbergen.

Growth rates of these new clones appear to be somewhat higher than those of the old clones already in use in the UK. They also have high resistance to diseases such as bacterial canker caused by *Xanthomonas populi*, *Melampsora* rust and poplar leaf spot caused by *Marssonina brunnea*. The genetic diversity of these eight clones is limited because several are full siblings, namely 'Primo' and 'Ghoy', 'Gaver' and 'Gibecq' and 'Beaupré' and 'Boelare'. However, another 50 clones from Belgium are being assessed by the Forestry Commission with a view to releasing selected clones during the next few years.[2] Paul Tabbush is, at the time of writing, preparing a book on short-rotation coppice for the Forestry Authority which will provide a lot more useful information for farmers and land-owners.

SIZE OF CUTTING

In Sweden they plant cuttings 20 cm long, but in the UK we use 25 cm cuttings — for no particular reason, according to Rod Parfitt. Murray Carter says survival is better with the longer cuttings, although they are not as easy to plant, and he advocates 20 cm for willow. The minimum thickness of a cutting may be around 7 mm. 'If they're thinner,' says Rod Parfitt, 'they just don't have the reserves to withstand a period of drought.' Thicker than about 2 cm gives no extra resilience but total volume of the cutting seems to be what's important. 'Long thin cuttings are OK — so are short fat ones — but not short thin ones,' says Rod.

The harvesting period for cuttings is between December and late February, before the sap starts rising. Cuttings are then stockpiled in cold store. At Long Ashton they keep the cuttings at just above 0 degrees Celsius. In Sweden they store at minus 2 to minus 4 degrees Celsius, which requires more sophisticated equipment but is more efficient. However, they store both small and very large boxes — some 4 feet by 3 feet square — and there are so many cuttings in each box it is difficult to cool them all down, so the boxes are ventilated, which causes a certain amount of drying. Consequently it is important to soak the cuttings before planting. At Long Ashton they store in plastic bags — but in small numbers.

When buying ready-cut cuttings they will probably be painted with a bright colour at one end. This has two purposes: first, to help the supplier identify the type of clone; and second, to show the person planting the cutting which is the top. It's not always possible to see which way the buds are pointing, at a glance, and they must not be planted upside down! Cuttings should not be allowed to dry out before planting, so soaking them upright in 10–15 cm of water for one or two days before planting will be beneficial and will stimulate root formation. The cuttings should be removed from the water and planted before the roots (which can be detected as little bumps under the bark) break through the bark or they may be broken off during planting.[3] When the weather warms up in the spring rooting begins within 24 hours of the cutting being immersed in water.

THE RUSH FOR CUTTINGS

If home-grown energy takes off as fast as expected there will be a shortage of planting material and this will attract all kinds and qualities of cutting on to the market. Imports will certainly be necessary; and with biomass industries based on coppice already established in Sweden, and to a small extent in Denmark, cuttings will need to be imported to supplement UK supplies. ETSU gives an example of the kind of quantities we are talking about: in order to service a 5 megawatt NFFO electricity plant using gasification at 25 per cent efficiency an output of about 2555 hectares of coppice will be needed, assuming an 11 dry tonnes per hectare per year yield. To be able to harvest a third of the area each year 850 ha per year for three years will need to be planted. In year one, then, 8.5 million cuttings will be required. So far, the UK government has not given farmers sufficient confidence that the industry will receive real pump-priming incentives to enable them to begin the dramatic planting schemes which would be necessary to meet the demand that could come when the result of NFFO applications have been received. If power stations fuelled by willow and poplar are indeed built in Cornwall, Hampshire, Suffolk, Northamptonshire and the Borders, then demand for planting material will be phenomenal.

WHERE TO GET PLANTING MATERIAL

Nevertheless, cuttings producers say they are ready for any such plantings. Murray Carter maintains his business alone could supply 'tens of millions of cuttings' from Yorkshire. The five demonstration farms across southern England are all potential sources of both willow and poplar planting material. (You will find their names and addresses at the back of this book.) If, though, you want to get hold of very large numbers of cuttings you may have to approach one of the specialist producers. As well as Murray and the five demonstration farms there are another three producers who specialise in providing cuttings —

Murray Carter says he can supply 'tens of millions of cuttings from Yorkshire'. (Clare Arron)

they all have trials on their farms and are in a position to provide large numbers of cuttings and give authoritative advice on establishing the crop. They are Hugh Snell of the Poplar Tree Company, Christopher Whinney who farms at North Molton in Devon and Lionel Hill who farms in Worcestershire.

The Poplar Tree Company is not primarily a producer of poplars for energy. The energy crop is just one of many by-products of this poplar-growing scheme whose end-product is hardwood timber. The company now has nearly 250 hectares of poplar using a new system of closely spaced planting, the new, high-yielding clones being tended in a particular way on just the right soils in joint ventures with 17 farmers in five counties. The dense planting allows early selection of the best trees; and there is earlier thinning for many different purposes.

Planting poplar under the auspices of the Forestry Authority's Woodland Grant Scheme (or MAFF's Farm Woodland Premium Scheme, which does not apply to short-rotation coppice) will require varieties (clones) which are on the authority's approved list. The authority is currently drawing up a recommended list of willow clones.

GROUP PLANTING

Where groups of farmers are working together, all planting willow to supply a local generating unit or heating boiler, it may pay them to include smallish mono-clonal plots on each farm. If planting material is taken from a mixed plot, some of the less vigorous clones might not come through and in time the mixture will change. Rather, each farmer could have five, six or even more plots, from which he or she could harvest planting material to make up mixtures of clones for production areas at home and on other group members' farms. Please note — there's much more to taking cuttings and successfully storing them than meets the eye. Gappy fields resulting from amateur cutting-takers bear evidence of this!

Malcolm Dawson of the Northern Ireland Department of Agriculture reminds us, though, that this would involve plant breeders' rights. If you have bought your planting material from a commercial supplier you must not sell cuttings without consulting your original supplier; and in the case of poplar, the beds from which cuttings are taken must be registered by the Forestry Authority. If your group decides to swap clones without charge it will still be safer to consult your supplier because, if you do not, you may find yourself excluded by your supplier and cut off from the supply of new planting material which you will need in future years. It may be that, to stay up with the very latest clones, it will pay, in the long run, to spend money on new planting material from commercial suppliers only. They will be providing much more than just planting material, once a good relationship is established: their advice on husbandry and crop protection will be invaluable. Breeders say there is a real danger that if they are unable to reap benefit from their work — as farmers 'grow their own', the breeders will revert to more profitable crops and desert short-rotation coppice crops — the required new strains and disease-tolerant types will not materialise. As the volume builds up, however, prices will fall dramatically, making DIY cutting production increasingly less attractive. Reliability and new genetic material will pay dividends.

SUMMARY — CHEAP WAY TO DISASTER

Planting low quality, low potential cuttings will lead to long-term disaster. Planting high quality foreign material which has not been trialled in the UK could also be extremely risky, because growing conditions vary with climate and latitude. Establishing a plantation of willow or poplar is like starting a pedigree herd of extra-long-life cattle. Choosing the first components of that herd must be done with the same great care that needs to be taken with any new project. The early rush for planting material may mean paying a premium for quality, and although prices will fall they should stabilise at a level at which breeders can afford to keep ahead of disease and the current needs of the industry, and at which multipliers can afford to harvest, process, store and supply top quality material.

References

1   Murray Carter, 'Willow breeding — new developments', paper at conference on 'Coppice — looking beyond set-aside' at RASE, NAC, 25 October 1994.
2   C. P. Mitchell, J. B. Ford-Robertson and M. P. Watters, 'Establishment and monitoring of large-scale trials of short-rotation coppice for energy', ETSU report B1255, 1993.
3   Murray Carter, at conference on 'Short-rotation coppice — growing for profit', NAC 24 March 1993.

CHAPTER 6

Establishing the Crop

'It's trees, therefore it's forestry!'

'Maybe it is trees but you have to treat them like cabbages so it's horticulture!'

'I'm growing short-rotation coppice on prime arable land using arable techniques so it's agriculture!'

The old divisions between farmers, foresters and horticultural growers, perpetuated by civil servants both in the UK and overseas, have rarely encouraged lateral thinking between these disciplines. The cultivation of short-rotation coppice, however, cuts across all three *and* brings in heat and power engineers too, who are the users of the product: it is they who will spell out what chip size they need; what the permitted variations will be; and what moisture content can be allowed. If you are a land-owner or forester who does not have practical experience of arable farming, swallow your pride and consult the best horti-culturalist or arable farmer locally. When he or she sees how good your coppice is looking next year, you may well be able to return the favour when it comes to the care of young trees. On the other hand, when designing a plantation, an arable or grassland farmer may need to seek advice from growers who have done this kind of thing before.

There's a lot to short-rotation coppice. As well as choosing the site best suited for the farm — perhaps on the wetter parts, although bearing in mind the hazard of coppice blocking the drains — the site must also suit the crop and the local environ-ment. It would be madness to grow the crop across someone's favourite view or in a position that would be visually inappropriate. It is vital, too, that the crop be easily accessible for harvest during the winter. If tractors and trailers or lorries might get bogged down every time they go to collect wood-chips, the site is wrong. It is the headlands that will cause the problems, according to coppice growers: between the rows the lacework of roots below the surface appears to support even heavy equipment such as the Swedish version of a sugar-cane harvester that has been demonstrated in this country for coppice harvest. The 'rides' to allow tractors, sprayers, fertiliser distributors and trailers easy access need to be carefully thought out as well, because they can take up between 10 and 20 per cent of the field. This unplanted area has great potential as a wildlife habitat, particularly the edges between coppice and ride. The Forestry Authority has an excellent leaflet about how to site 'Short-rotation coppice in the landscape'. It is Forestry Practice Advice Note no. 1, and you can obtain it free from the Forestry Commision in Edinburgh.

To be eligible for a Forestry Authority Farm Woodland Grant Scheme grant you must submit a management plan which will satisfy environmental conditions set by national organisations. These conditions will affect the design of plantations and the sites upon which they may be located.[1] The planting schedule will depend on the chosen rotation. You will need to divide the land to be put under short-rotation coppice into similar areas and plant them in successive years. Use the best advice you can get! It may cost money, but your supplier of cuttings ought to be experienced in the good layout and design of coppice plantation, especially if he or she offers a complete planting service, including cuttings, fencing and mechanised planting. Alternatively, if you are a grassland farmer and are applying for a woodland grant, the local Forestry Authority representative will be able to help. ETSU has established good working relationships with many environmental organisations and Caroline Foster at ETSU will be able to offer useful advice on how to fit coppice uneventfully and satisfactorily into the landscape.

(text continues on page 83)

Short-rotation coppice in the landscape — some 'do's and don'ts'. (Taken from the Forestry Authority's excellent *Forestry Practice Advice Note 1*, obtainable from Forestry Commission Publications Department; see Chapter 15 for the address.)

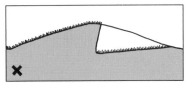

Coupe splits hill and conflicts with landform: poor.

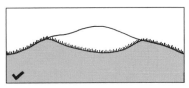

Coupe forms hill top cap and follows landform: good.

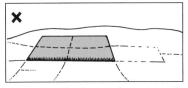

Geometry too dominant and contrast with field patterns too strong.

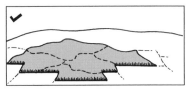

Coppice area relates to landform on upper slope, fits into field pattern at lower margin. Coupes follow landform, not old field pattern.

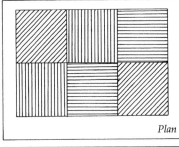

Plan

Perspective

Rectangular, geometric layout.

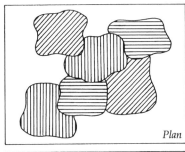

Plan

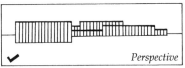

Perspective

Irregular, interlocking layout using the varied heights to good effect.

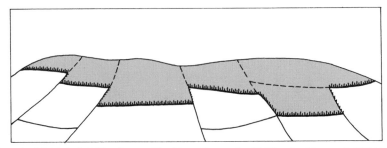

Layout related to both field pattern and landform. Coupes should follow landform, not old fields.

To avoid the effect of parallel strips it is advisable to plan an interlocking pattern. Row directions can alter and extraction routes can be accommodated between the coupes.

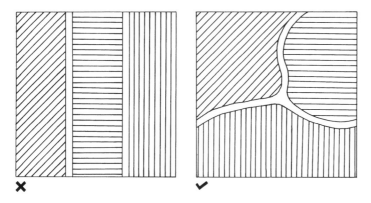

The internal shape of a coppice coupe:

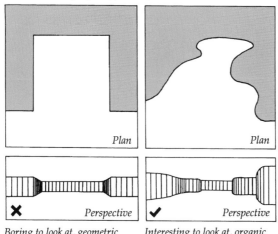

Plan

Plan

Perspective

Perspective

Boring to look at, geometric form dominates

Interesting to look at, organic form allows space to flow

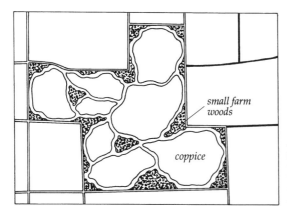

small farm woods

coppice

In this example larger areas of coppice have been interspersed with smaller woodland clumps intended to remain as more permanent features.

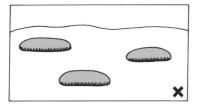

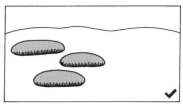

Three isolated blocks float in the landscape and are out of scale

Better – cluster them near each other, but size is too similar.

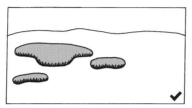

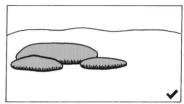

Vary size and position them near each other.

Make them appear to be parts of one whole to improve scale.

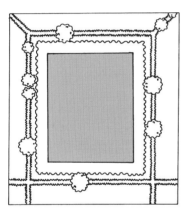

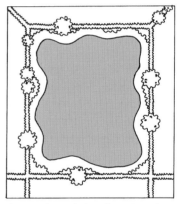

Belt of trees hides coppice but does not blend into grain of landscape.

Groups of trees fit into the grain and texture of the landscape.

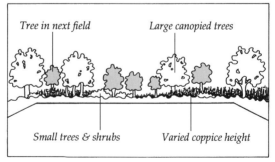

A view towards a well designed coppice layout shows how the varied heights and groups of trees tie the composition together.

TIMING OF PLANTING

Planting of short-rotation coppice takes place in winter or early spring and this is the stage during which foresters have to accept the fact that they must treat it as an arable crop. Cultivations and weed control prior to planting are vitally important in establishing a strong stand of coppice. It is not enough simply to push cuttings into a mess of weeds and then come back in six months' time to see whether any survived. Checking the soil to find out its nutrient status is also essential. You may have already done this when considering the most appropriate site for the crop, but the advice is to take soil samples across the whole site, to check nutrient status, acidity (pH) and soil type. Soil samples can be sent to ADAS (Agricultural Development and Advisory Service), agricultural colleges or private laboratories. Once you get the results you may need to apply lime or fertiliser to correct deficiencies — although, as we see in Chapter 8, short-rotation coppice requires much less fertiliser than, for example, cereals. Tackling particular weeds before cultivation may be necessary.

Weeds like couch grass will make life a misery if not controlled early on. The easiest way to tackle this, and various other perennial weeds, may be to use a complete herbicide such as glyphosate while the weeds are still growing strongly, before they die down for winter. If coppice is to follow a cereal crop, then pre-harvest systemic herbicide spraying may be advisable. Again, it will pay to use the best-qualified local knowledge available about how to control specific weeds on your particular soils (see also Chapter 7, concerning crop protection). If cold storage is available for the cuttings, then you could catch the weeds as they become vigorous again in spring. Hold the cuttings back in the store, at a temperature of minus 2 to minus 4 degrees Celsius,[2] to prevent them sprouting — if they do sprout they could be damaged during planting — and delay planting until you have sprayed off the weeds in the spring using glyphosate, paraquat or something else appropriate to your weeds. The dead weeds can form a partial mulch and a residual herbicide can then be applied for even more complete weed control. This will, of course, shorten the growing season for the coppice. Organic farmers will have to develop their own ways of weed control, although if the planting is to take place on land which has been well farmed organically for several years, there should be no particular problem. The organic solution to weed control may be to opt for planting through a mulch, or some kind of pre- and post-planting cultivation and inter-row weeding.

TILTH AND PLANTING

As with any arable crop, there are several reasons for cultivation before planting. Cultivation for weed control is of major importance, but the soil also has to be prepared so that it can provide water, nutrients and oxygen to allow root development of the crop. For many sites deep (25–35 cm) ploughing followed by conventional cultivations may be sufficient, although this may need to be preceded by sub-soiling if there is any plough pan or compacted layer which could inhibit root penetration. All the old rules about cultivation have to be followed — carry out the work when the weather and soil conditions allow it, for example harrowing will only give a good tilth if it's done when the soil is fairly dry, otherwise it will not break down into smaller fragments.[3] The aim is to finish up with a clean tilth similar to

Planting willow cuttings into 'a good friable tilth' in Sweden. (Stig Ledin)

that needed for sowing cereals or forage maize — a 'good friable tilth' — but it should be deep enough to accept 20–25 cm cuttings. Dr Paul Maryan, in his paper to the 1993 ETSU conference about short-rotation coppice, suggested that ground preparation should be 'as for any root vegetable'.

Bad weather may delay planting in the spring, but if a cold store is available for the cuttings the wait will be worthwhile because, when they are finally planted, there will be better contact between the cutting stems and the soil; moisture will be more readily available for the cuttings and this will encourage roots to shoot. If the soil is 'forced', to allow planting by a certain date, it may not break down sufficiently and contact between cutting and soil will be poor, allowing cuttings to dry out, which can be fatal. Drying out, in fact, is one of the major dangers facing a willow or poplar cutting and a dry first summer can set the crop back severely on some soils if no irrigation is available. Yorkshire willow breeder and grower Murray Carter maintains that good weed control is the key to conserving moisture in the soil – it is more important than irrigation. There are, however, legendary tales of cuttings surviving under the most horrendous conditions, even though the resulting crop will not have the productive potential of one which has had an ideal seedbed, weed control, plenty of moisture and no checks to its growth.

Lionel Hill, the cuttings producer at Feckenham in Worcestershire, tells the story of how he wanted to experiment with poplar coppice on landfill sites and spoil tips. He had a plot on to which had been spread sub-soil clay and he was ready to plant the pencil-sized poplar cuttings. Ploughs and cultivators would not touch it — they just bounced off — so Lionel tried an iron bar to make a hole for each cutting, without any success whatever. His intrepid planting team then went down on hands and knees with hammer and cold chisel, but still failed to penetrate the hard clay. Finally Lionel got hold of battery-driven hand drills and drilled holes just large enough for each cutting to be slid into. It was a remarkable success and the take has been almost complete. The trees are now between 5 and 12 feet high, and thriving — but there is hardly a weed in sight on the bare hard clay surface. Like Lionel, you could plant your coppice by hand, but unless it is an experimental plot it will prove very expensive!

According to Murray Carter we can expect coppice plant-ing to build up sharply from 1995 to 2010, to keep up with the projected demand for home-grown energy. He says that although, in theory, the planting season extends from December to late May (if cold storage is available), in reality, due to bad weather, adverse soil conditions and other winter work, you may finish up with only about 60 days a year, discounting weekends, in which to plant a massive area . That will need a planting rate of hundreds of hectares per day and to achieve this will take a high degree of mechanisation.

MACHINES FOR PLANTING

Murray Carter has a 'step planting' machine from Salix Mas-kiner AB, the Swedish manufacturer, and he is their agent in the UK. The machine is called a step planter because each planting mechanism moves in steps, so that while a cutting is being pressed into the soil the mechanism is not moving forward. It plants four rows of cuttings at once at the amazing speed of eight cuttings per second. In practice, Murray says the machine can plant 1–1.5 hectares per hour, where 10,000 cuttings per hectare are required. A tractor driver and a machine operator are needed for each machine. The machine uses whole rods of willow, up to 2 metres long, chopping them off just before they are pressed down, almost out of sight, into the ground. The cuttings finish up vertically in the ground and can then be rolled to consolidate them and to help conserve moisture. The rolling just presses them a little further into the soil.[4]

The Maskiner machine costs £27,000 but Murray Carter says that, compared with using two conventional planters which were based on cabbage planting machines, each needing four operators to give the same rate of work, the machine soon pays for itself in saving labour costs. In a direct cost comparison for planting a comparatively small area of 250 hectares between the conventional four-row planter and the new 'step' machine, it would cost £159/ha with the old planter and £124/ha with the new one. Over a larger planted area, however, say 500 ha, it would cost only £70/ha with the new planter. Don't expect your contractor to offer that price, though, because it does not include a profit margin or overhead expenses!

Salix Maskiner AB's 'step planting' machine at work in Sweden. It cuts off the right length of rod and plants it vertically 'up to the hilt', at a surprisingly high rate per hour. (Stig Ledin)

Another Swedish machine is the Fröbbesta: a two-row model fits on the three-point hydraulic linkage behind a tractor and two operators feed cuttings, one at a time, down a tube which descends behind two share-type coulters which cut grooves into the seedbed. They are dropped in time with a clicker, to achieve the correct spacing. The cuttings do not simply fall down the feed chutes but are fired down by a pair of hydraulically driven counter-rotating wheels at the top of each chute. A development of this planter has an automatic feed hopper: with this machine only a driver and one operator are needed for a two-row machine, with the machine operator topping up the hopper. This improves the work rate from 0.5 ha per hour to 1.1 ha per hour.[5]

A two-row Fröbbesta planter at work in Sweden. (Stig Ledin)

STARTING SMALL

Most people will start by planting just a small plot of coppice, to get to know the clones and the techniques for growing the crop. One of the five demonstration farmers, Robert Goodwin from Kelvedon in Essex, is developing a planter suitable for these small beginnings. On his two-row prototype cuttings are dropped manually into a feed mechanism that delivers them down a drop-tube at metered spacing. The machine will be completely automated eventually; the only bottleneck at present is the speed at which the operators can put cuttings into the feed mechanism. The target planting rate is one hectare in three

Robert Goodwin's planter.

hours. The aim is to plant the cutting vertically so that its top is level with the ground surface. Robert, too, believes in rolling after planting — although he advocates extra care when rolling poplar because the cuttings are more fragile than willow; if cuttings are not level with the ground they are likely to be snapped off. A commercial version of Robert Goodwin's machine should be on the market in 1995.

John Turton of Turton Engineering is developing a 'cut and plant' machine designed to plant 6000 cuttings per hour (0.6 ha/hour). It uses whole rods, cutting and planting in one repeated cycle. The basic unit has two rows, and requires one operator. A four-row unit would need two operators and a tractor driver. The machine plants in staggered rows with a spacing of 0.75 metres. The cut and plant coppice planter will be on the market in 1995. John Seed of Border Biofuels says that 'just about any vegetable planter can do the job', although when it comes to larger acreages he will be looking at some of the bigger, faster planters.

Planting by hand is certainly not out of the question but obviously takes time: an estimated 35 hours per hectare.[6] Willow

Neil Roberts and his assistant (of Murray Carter Horticulture) planting a trial plot by hand in Wiltshire.

and poplar cuttings must be planted 'the right way up', i.e. the way they were originally growing on their mother plant.

PLANT POPULATION AND SPACING

For the United Kingdom, the recommended number of plants per hectare is 10,000 for the moment.[7] John Seed of Border Biofuels, however, talks about 10–15,000 per hectare, while scientists at Long Ashton discuss 15–20,000 per hectare to give denser cover and to maximise yield especially in the early years.[8] John Seed's argument is that if fast green cover is desired, then denser spacing will give you that by intercepting as much solar energy as possible and converting it into biomass. 'It's a trade-off between green cover and the cost of the cuttings,' he tells me. 'Some will die out in years two and three, and eventually the willow will "decide its own density" as more dominant plants smother weaker ones.' In Sweden they plant 18,000 cuttings per hectare. Poplar may be planted less densely if a longer rotation is being used. It can be harvested at intervals of between two and five years, and this can be repeated at similar intervals. For the five-year rotation, density can come down to 6700 trees per hectare, which might suit areas where existing forestry equipment and expertise is available. Yield should not differ.[9] Paul Tabbush of the Forestry Commission has some interesting views on what might be considered when it comes to planting poplar, with different plant populations for different areas of a plantation (see Chapter 5).

The cutting is planted 'up to the hilt'. That's how it should look — except that the tilth in this field is not as fine as it should be. (The reason: late planting after a wet period on heavy land. We all have our excuses!) (George Macpherson)

John Seed believes that a lot more research is needed to discover the right density of planting for any particular soil type or situation. In the absence of scientific findings, he says, growers will have to make their own trials. There are several schools of thought, too, about the ideal spacing for short-rotation coppice. Each cutting needs enough room to be able to produce optimal yields of biomass. At the same time it is vital to be able to harvest the crop without running over the plants — so room has to be left for tyres, or tracks to pass, where necessary. The latest recommendation from Damian Culshaw, who is the agricultural engineer at ETSU, following discussions around the UK industry, is to plant in twin rows with 75 cm (0.75 m) between rows and the two rows staggered so that plants are not opposite each other, but opposite the space between plants in the next row. Cuttings in the row are spaced at 90 cm (0.9 m) apart, as shown in the diagram.

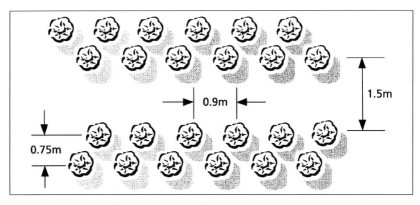

How to space double rows of cuttings for short-rotation coppice. This is taken from ETSU's information newsletter, *Wood Fuel Now!*, issue no. 3, October 1993.

The space between the pairs of twin rows should be 1.5 m, to allow for variability in track and tyre width, including balloon tyres. A few weeks after planting it will pay to go round and replace any cuttings which have not sprouted but have obviously died. This is not a cheap operation and may take five hours per hectare. Another operation that may be necessary — depending on where you live and the local population of rabbits or deer — is fencing. This may be carried out by the planting contractor as part of the establishment package. We look at fencing in Chapter 7.

CUTTING BACK

At the end of the first season, most growers cut their willow or poplar to within a few centimetres of the ground. This is to stimulate the plants to send out many shoots: that's the 'coppicing' effect. This practice may change, however, and a few growers are talking about letting the plants grow on for a second or even a third year before the first cutback — although this leads to a 'thinner' stand of willow, since there are more single stems. Murray Carter has tried this and discovered that, on his land, the thinner crop did not smother the weeds nearly as well and he 'ran into big weed problems'.[10] Using a higher planting density may produce commercially worthwhile yields. One argument for a delayed first cut is that it allows the plant to become better established before hitting it so hard. It also allows an earlier first harvest, even if that harvest is not as large. This harvest can provide either more planting material, or wood-chips. Using a rod harvester would give you planting material, but if the stand is a mixed one there is a possibility that, if this is repeated, some clones might come to dominate new stands because they produce more stems, reducing the benefits of genetic diversity in the stands.

The cutback can be done with a brush cutter and the rods collected, or left to rot down. The rods could possibly be harvested for a light crop of woodchips, using a forage har-vester. I am told that John Seed has successfully tried a rape swather for the cutting back. At the same time as cutting back, consider how you will fill the gaps left by the odd failure of cuttings to take. Do not use new, 20 cm length cuttings if you want the plant to catch up its companions, which by now are well established.

Neil Roberts, who works with Murray Carter in north York-shire, gave me some advice. 'Use rods from 50 cm to 1 metre long,' he said. That would give them a better start, and they would catch up after a year or two. Another important hint: 'Don't go gapping up (or beating up as some people call it) until *after* you have applied the herbicide — if you are going to use it in the spring,' he said. 'Residuals will damage the cuttings if sprayed on after gapping up.' That gave me a problem, because it would mean storing cuttings, made in December or January, in cool store until late February or March. Neil solved that one, too. 'Leave them on the stool (plant) until after you have sprayed,' he said.

After the first cutback, the harvest will not take place again until between two and five years, depending on which rotation has been planned. In the first year after cutting back regrowth is dense and vigorous, with rods reaching 3–4 metres in height by the end of the season. Current expectations are that the planta-tion will be harvested at least another eight times before too many of the stools die out.

References

1   J. Ford-Robertson, P. Mitchell and M. Watters, Handbook on how to grow short-rotation forests, Swedish University of Agricultural Sciences, 1992.

2   S. Ledin, 'The development of short-rotation forestry in Sweden', RASE and WEDG conference at NAC, 1993.

3   See note 1.

4   H. Johansson, video, 'Salix growing energy', Swedish University of Agricultural Sciences, 1993.

5   Damian Culshaw, 'Status of short-rotation forestry mechanisation worldwide', IEA Report, ETSU, 1993.

6   Centre for Agricultural Strategy, 'Energy coppice: an alternative farm crop?', issue no. 2, 1991.

7   Damian Culshaw, 'Coppice row spacing — a recommendation', *Wood Fuel Now!*, Issue no. 3, October 1993.

8   R. Parfitt, G. Hunt, E. Thompson and T. Badmington, 'Wood production from short-rotation coppice, Mercia programme', ETSU, 1994.

9   ETSU, 'An assessment of renewable energy for the UK', HMSO, 1994, pp. 245–6.

10 Personal communication, 1994.

CHAPTER 7

Protecting the Crop

Willow is the sort of crop that gives you encouragement. It can be in leaf a week after planting, given the right weather and soil conditions. The pencil-sized cuttings, pressed down almost out of sight into the soil, send up bright green shoots, several at a time. They are tender, though, and this is the time when they need protection if they are to become established strongly enough to produce profit for the next 30 years. Poplar cuttings are not quite as quick, taking perhaps three to four weeks to send out shoots, depending on the time of year. This is a time for people with forestry experience to become arable minded! It's no good going away for a couple of months, then returning to see 'whether they've taken or not'. You must be there at the end of the first week to see whether the rabbits are taking out

The first tender shoots (highly prized by slugs and other predators) appear after just a few days.

those first tender shoots. If they are, the rabbits have to be stopped: the young shoots *must* be protected. We shall discuss rabbits and their control later in this chapter, along with other enemies of the crop. Chief among these is the disease rust, which looms as a lethal threat to the future of short-rotation coppiced willow and poplar in a maritime climate such as the United Kingdom.

Less than a month's growth — even though the tilth was too cloddy. This was before the drought!

The other major hazard at this time comes from weeds. There must not be a weed in sight, because weeds take the moisture which is absolutely essential for the well-being of the unrooted coppice cuttings. Weed control should have started before the coppice was planted (see Chapter 6), but now it must be reinforced. Couch grass, for example, should be allowed to grow sufficient leaf so it can be treated with a herbicide like glyphosate, using a spot weed wiper. Do not use a knapsack sprayer. According to Neil Roberts, manager for Murray Carter, even a tiny amount of glyphosate drift could severely endanger young willow at this stage. He says 'take local advice' about using something like Fusilade.

WEEDS WILL HALVE YOUR YIELDS

Uncontrolled weeds in the first season will reduce coppice growth by 50 per cent, and dry matter yield to 20 per cent of that achieved by willows or poplars in weed-free conditions, according to work done by Rod Parfitt and others in the early 1990s. They recommend that within a few days of planting the site should be over-sprayed with a residual herbicide mixture. This would be to control weed seeds as they germinate (look at the previous season to see which weeds are prevalent) without damaging the newly planted cuttings. Ideally, the application of residual herbicide should be done after rain has consolidated soil around the cuttings, but before shoot or weed emergence.

A three-year programme at Long Ashton funded by ETSU showed that various combinations of metazachlor, pen-dimethalin, propyzamide, metamitron and simazine were both safe and effective.[1] Any weeds that develop despite all this should be hit with carefully directed spot treatments of foliar-acting herbicide or inter-row cultivation. If any weeds are still around when it comes to cutting back at the end of the first season, work done by Rod Parfitt at Long Ashton in the late 1980s showed that the coppice stools can be over-sprayed while still dormant, or just resprouting, with a foliar-acting herbicide such as amitrole or clopyralid, with a further residual treatment to maintain weed-free conditions through the harvest cycle. On organic soils, as usual when using herbicides, special problems have to be taken into consideration because of chemical interactions in the soil.

While some publicity has been given to short-rotation coppice needing herbicide treatment only once during its 30-year life, Murray Carter and other growers would disagree. 'You need to control the weeds every time you harvest the crop,' says Murray. He takes visitors to see a plot where willow had been harvested and where it had failed to regenerate strongly because of weed competition, even though the root system had been well established. This plot — mercifully small — was a disaster. Complete freedom from weeds both at planting and after harvest cannot be too strongly emphasised, he says, and he is backed up by Malcolm Dawson of the Department of Agriculture at Loughgall in Northern Ireland. He told me that, as a

A 'finger-weeder' for control of the new crop of weeds, which emerge in July, in Sweden. (Stig Ledin)

Finger-weeding in progress. (Stig Ledin)

trial, he had left a plot of willow without weed control after harvest — to see whether it was really necessary. 'It turned out to be a real mess,' he said. 'In our climate, where quite often weeds grow all winter while willow is dormant, the weeds get a head start in the growing season next spring.' Malcolm says that even if there are just a few weeds present in the spring in a newly harvested crop, but particularly for a new crop, 'knock them over'. Dr Stig Ledin, one of Sweden's top specialists in short-rotation coppice, agrees: 'During the entire first year, the *Salix* plantation is extremely weak in its competition with the weeds. When it has become established and can effectively shade the soil, then the weeds are a minor problem.'[2] In Sweden, when a new crop of weeds emerges in July, they employ mechanical weed control. Harrows, rotary cultivators or cultivators with spaced tines can be used, with the tractor straddling a double row.

THE NON-CHEMICAL APPROACH

A particular advantage of short-rotation forestry crops is that they are less sensitive to mechanical injury than shallow-rooted crops. A garden-type rototiller may be handy, working at 15–20 cm deep, but often you may need to go in with a hand hoe for the odd stubborn weed. Black polythene mulch has also been tested for weed control in short-rotation coppice in Canada. When tested against chemical and mechanical controls it was the only treatment that actually increased plant productivity on a poorly drained site.[3] David Clay of Avon Vegetation Research, however, is not as confident about the ability of willow to tolerate being bashed about by inter-row cultivations from June onwards in the first year: he believes this may cause severe damage to willow roots and shoots. There's also the problem of weeds between the plants in the row which will not be touched by the cultivator.

Organic farmers will be pleased to note that the polythene mulch helped regulate soil temperature and moisture (in Canada). It is always 2–4 degrees Celsius higher under the plastic than below the surface of bare soil, and humidity is always slightly higher during periods of drought while remaining lower after heavy rains. However good it is, though, plastic mulch is not cheap, with both labour and materials being costly. At ETSU they have found that if this mulch rips, or is not securely held down, the flapping plastic can knock off tender buds and prevent sprouting.

I have not found anyone who has used a mulch of woodchips. It may be worth consideration, although David Clay thinks organic mulches are unlikely to succeed for agronomic and cost reasons.[4] Some growers think it may be too expensive, with the markets for woodchips increasing at such a pace! David Clay believes that attention to weed control must be continued for the first two years of the crop's life. The crop rarely recovers from the damage caused by early weed smothering, he says: unrestricted growth is all-important. If cuttings are smothered there is permanent loss because replacement cuttings never catch up with the rest of the crop. Weeds in late summer may not be so competitive but they are still dangerous because their seeds will germinate in the following season and cause problems. He maintains that the weed control expertise needed for establishing short-rotation coppice is more akin to that of the horticulturalist because the planting material is small and non-competitive.

Not only does weed control have to be pre-planned, but instant action must be taken when weeds escape due to adverse weather or unexpected weed resistance. David Clay's report for ETSU on weed control in short-rotation coppice contains useful lists of weeds and herbicides and plenty of good advice.[5] After cutting back, at the end of the first year, the crop needs weed protection at least until June or July of the second year, when the leaf canopy of the crop closes across the rows again. Once this takes place the crop should be able to look after itself until after harvest two or three years later — except where there are vigorous climbing weeds like bindweed. Timing the application of a mixture of contact and residual herbicide after cutting back is critical: spraying in cold weather may not be as effective on the weeds, but waiting until coppice shoots appear may lead to a check in crop growth. Be wary, though, because it is easy to spend a fortune on weed control during the establishment of short-rotation coppice: pre-planning is absolutely essential.

Beware also the regulations controlling the use of herbicides. It is fine to use products with label recommendations for forestry, but few of these are appropriate for overall spraying in the crop. The Forestry Commission has obtained off-label approval for a number of products for farm forestry use, which permits spraying in a particular situation where all the conditions for use on the label are followed. It's all at your own risk, though, regarding efficacy and safety. Take advice: firms such as Banks of Sandy

HERBICIDES WHICH HAVE BEEN TRIED ON SRC PLANTATIONS

Active ingredient	Product name	Type of activity	Use situations
Amitrole	Weedazol TL[a]	Foliar-acting	Overall pre-planting Directed between rows Overall after cutting back willow
Asulam	Asulox	Foliar-acting	Directed spray onto weeds
Clopyralid	Dow Shield[b]	Foliar-acting	Overall or directed spray onto emerged weeds
Cyanazine	Fortrol[b]	Foliar and soil-acting	Overall post-planting spray
Fluazifop-P-butyl	Fusilade 5[b]	Foliar-acting	Overall application onto emerged grass weeds
Glufosinate	Challenge	Foliar-acting	Overall pre-planting Directed spray onto emerged weeds
Glyphosates	(various)	Foliar-acting	Overall pre-planting Carefully directed spray onto emerged weeds
Isoxaben	Flexidor 125[b]	Soil-acting	Overall post-planting spray
Metazachlor	Butisan S[b]	Soil-acting	Overall post-planting spray
Paraquat	Gramoxone 100	Foliar-acting	Overall pre-planting. Directed spray onto emerged weeds
Pendimethalin	Stomp 400[b]	Soil-acting	Overall post-planting spray
Propyzamide	Kerb	Soil-acting	Winter treatment as overall spray or granules, 9 months post-planting or later
Simazine	(various)[c]	Soil-acting	Overall post-planting spray Overall after first year cut-back
Triclopyr	Timbrel	Foliar-acting	Carefully directed spray onto emerged weeds

[a] Off-label Approval applied for   [b] Off-label Approval granted
[c] Off-label Approval has been applied for, for 'Unicrop Flowable Simazine'.
Off-label Approval granted for specific products.

The above table is based on small-scale experimentation and limited commercial use.

The herbicides mentioned have been approved by MAFF for use in SRC and the terms of the approval should be strictly adhered to. Before any product is applied in coppice, users should obtain copies of the Approval Notice from MAFF which lays down specific conditions for use. Application of products with Off-label Approval for coppice is entirely at the commercial risk of the user.

All pesticides in the UK must comply with the terms of approval under the Control of Pesticides Regulations (COPR). Farmers and growers must take all reasonable precautions when using pesticides. Advice on how they can meet their responsibilities under this legislation is given in the Code of Practice for the Safe Use of Pesticides on Farms and Holdings. Copies can be obtained from HMSO (ISBN 0-11-885673-1), price £5.00.

An extract from Agriculture and Forestry fact sheet no. 5, from the DTI. (The information was amended in July 1994 and was accurate to the best of the DTI's, ETSU's and the author's knowledge at the time of printing. No liability can be accepted for its use: it is wise to take professional advice before using herbicides on short-rotation coppice.)

and Doltons are offering a complete agronomy package for those farmers and landowners contracting with electricity generation companies. This package also includes initial clone selection and plant protection throughout the crop's life.[6]

INSECTS

Both willow and poplar short-rotation coppice may be attacked by a wide range of leaf-eating, stem-sucking and wood-boring insect pests: that's partly what makes the crop so attractive a habitat for so many different species of birds. It is said that willow plays host and hostess to more insect life than almost any other crop but, according to scientists at Long Ashton, few of these insects have so far become a serious problem in mature stands of coppice, with the exception of the brassy willow beetle. Natural predators and the huge number of leaves should mitigate against this.

When new shoots in a young stand or after cutback or harvest are being badly attacked (by such pests as leaf or brassy beetles) it may be necessary to spray with an insecticide to prevent a potentially devastating population explosion. Slugs, too, are an early invader of willow. As soon as the young green shoots emerge from the cuttings, slugs move in to nip them off: they need to be controlled if setbacks for the crop are to be avoided. Leatherjackets can also be a serious threat, particularly where cuttings are planted on ploughed-in grassland.[7] The important thing is to be aware that such a possibility exists and to keep a close eye on the crop during vulnerable periods, seeking advice where necessary. An attractive idea for insect control is to allow free-range poultry or game birds access to young stands of coppice, which would both reduce weeds and keep insects and invertebrate pests down. This might even reduce coppice costs by bringing in extra income from the livestock involved — an argument, perhaps, for calling short-rotation coppice 'agroforestry'.

RABBITS AND DEER

Rabbits may not be a problem — for a few lucky people — but if there is any risk at all of rabbit damage it would be foolish not

Note the rabbit fencing in the foreground. The garden rotovator worked wonders on this trial plot. The spacing does not conform to field scale short-rotation coppice but is at 75 cm between the rows and between the cuttings in the row.

to protect against it: the damage they cause is often terminal. It is also interesting to note that in some places some clones may be debarked by rabbits while other clones are left untouched. Conventional rabbit fencing, such as part-buried wire netting, is often used to protect short-rotation coppice but electric fencing is also available, both the low-level electrified mesh and the more robust multi-strand fencing, such as that made by Rappa, in Stockbridge, Hampshire. James Ridley of Rappa has been quoting some very competitive prices per metre for special rabbit fencing. To use this fencing you need fairly level ground — where a combine harvester can easily go. The fence should be installed at the same time as planting, with a strip of persistent herbicide in a band under it, to prevent shorting out through grass or other green material.

Rappa can supply fencing to keep out rabbits *and* deer, which may also be a problem. In trials to compare row widths in Sweden, researcher Eva Willebrand discovered that roe deer preferred wider spaced crops. They certainly grazed them more heavily. We should be grateful, however, that, unlike the Swedes, we do not have to cope with moose as well. Apparently they graze the willow when it is young, although when it grows out of their reach, to 3 metres, they only graze the edges of plantations.

Alick Barnes, with his poplar coppice in Devon, hopes to do without rabbit fencing. Instead, during the vulnerable establishment period he has been using Renardine dripped along the edge of the field every four days or so for the first six or seven weeks, with varying success. It cost him £60 in materials in one year for 24 hectares of coppice. No doubt he would have been visiting the crop every four days anyway, to watch progress, so labour costs could be considered as incidental. He says the rabbits just love his 'Boelare' poplar but are not so keen on the 'Beaupré'.[8] Guy Lee, who farms 400 acres near St Boswells in the Scottish Borders, has spent a good deal on rabbit protection, with dug-in netting. 'Our rabbit population is not too serious but by all reports even a few can ruin an otherwise successful crop,' he says.[9] In a clean field mice and field voles should be no problem, but in a field where there is a lot of withered and dead weed material they may use it for cover and come out to gnaw the stems.[10] Grey squirrels can cause a lot of damage to poplar by stripping the bark in the spring and early summer. The only effective control is to eliminate the grey squirrel population locally using warfarin-baited hoppers (inaccessible to other animals and birds) during the critical period between March and July.[11]

THE DREADED RUST DISEASE

Willow coppice has one major foe: *Melampsora*, or fungal rust. In the UK it reared its ugly head in 1986 when the *Salix burjatica* clone 'Korso', which had been grown for 30 years without showing any signs of the disease, suddenly suffered badly. That year there were outbreaks in the Republic of Ireland, Northern Ireland and in England. In the next few years attacks began earlier in the season and became more severe. At the same time, in New Zealand, *Salix viminalis* and its hybrids were also attacked repeatedly by what seemed to be a new form of the rust pathogen. By 1988 severe rust, leading to early defoliation, occurred on more than half of the 105 clone plantings recorded by Long Ashton Research Station.[12] 'Korso' has been so severely attacked ever since that it can no longer be grown productively in some areas. Even 'Bowles hybrid' was seen to be infected by rust in 1987 and this has remained a problem (although generally this is no longer seen to be of major importance if mixtures of clones are planted). The worst damage is caused with early outbreaks in the growing season, exacerbated by suitable weather conditions for rust development. 'Korso' regularly loses at least 30 per cent of its yield due to rust, although attacks later in the season may not be as devastating.

There is virtually no question of spraying against rust because the value of the crop does not warrant it. Spraying is also impractical, using any conventional machinery. The answer, for the present, would seem to be planting mixed clonal stands of coppice and staying ahead of the disease by breeding resistant or tolerant varieties. This accounts for the high degree of interest by breeders in Sweden and the UK to find such varieties. There is some hope that research at Long Ashton may discover a biological control system for rust, because it has been found that pustules of rust on stems of 'Bowles hybrid' are themselves infected by a parasitic fungus, *Sphaerellopsis filum*. This gets most of its nourishment by predating on the rust pathogen at the rust's expense.

For the past few years Dr David Royle and his colleagues at Long Ashton, who specialise in the diseases of willow and poplar, have been observing impressive natural control of rust in a mature plantation of the *Salix viminalis* cultivar 'Bowles hybrid' by this hyperparasite, without humans having anything to do with it! Rust does not usually affect the stems, but this clone is an exception. The fungal disease of rust overwinters in the rust stem cankers found on 'Bowles hybrid' and, as I mentioned in Chapter 5, it may be possible to encourage this fungus to fight rust naturally, within plantations. For example, in mixtures of willow clones you could include those which harbour the stem-infecting form of rust (like 'Bowles hybrid') which carry the biological control organism. In this way *Sphaerellopsis* might be encouraged to spread more rapidly, thus helping to suppress rust throughout the crop. You could even produce 'Bowles hybrid' clones for sale carrying the *Sphaerellopsis* organism to be included in willow mixtures, as a measure to keep *Melampsora* under control.

The importance of this possible control of rust has implications for the use of pesticides. If fungicide is used in an attempt to control rust, the beneficial fungus will be killed off as well. Long Ashton is advocating a national policy of avoiding the use of fungicides in short- rotation coppice, except when absolutely necessary in propagation nurseries.

Three species of *Melampsora* cause most of the rust disease on biomass willow in the UK. *Melampsora epitea* is the most widespread and causes the worst outbreaks. Its life-cycle involves both the willow and the European larch as host plants. Depending on the species, rust may appear in three different ways: as one or more foci of infection (in 'Korso' only) on the leaves of young shoots of the odd plant in May; as lightly infected leaves distributed throughout the plantation in May or June; or it may arise in April from over-wintered stem cankers, with fresh pustules on the cankers, or on developing lateral willow buds and shoots. The signs are typical yellowish-orange pustules. The disease spreads fast, depending on the weather. *Melampsora* varies between sites in its ability to attack willow, and by 1991 Long Ashton had detected eight different types or races of the disease which attack short-rotation coppice willow in the south of England. There are likely to be more, across Europe and North America. It may be of some comfort to learn that experience from surveys done by Long Ashton suggests that serious, yield-debilitating epidemics are uncommon on a wide range of clones, with only certain clones, such as 'Korso', succumbing regularly. Hence the importance of mixing clones in a plantation (as described in Chapter 5).

Poplars suffer from a similar but distinct range of diseases and are potentially at just as much risk. At present they do not have the same reputation as willow for susceptibility, partly because they have not so far been grown (to any large extent) in monoclonal stands for biomass production. Also, they have been subjected to far more intensive research and breeding for much longer than willows. In the USA, however, over the past few years, there have been devastating attacks of poplar rust — which indicates just what can happen when large, monoclonal stands replace the natural mixtures of trees which occur in nature.

In 1991 Long Ashton set up a network of field trials to examine the distribution of damaging rust types on different willow varieties in the UK and abroad. Through the International Energy Agency a standard set of 24 clones, 12 from the UK, 6 from Sweden and 6 from Canada were planted in a common design at Long Ashton, Loughgall (Northern Ireland) and Craibstone (Scotland), as well as in Sweden and Canada. Rust is now being monitored regularly and the results are being assessed.

RUST MANAGEMENT

Current thinking on rust management is to plant mixtures of clones. This should give long-term protection against the disease, but it must be remembered that you can't change the components of a mixture once it has been planted. Much will depend on a correct choice of mixtures at the outset. Long Ashton is working on recommendations as to the choice of mixtures. Some people are now wondering whether it might not be better to mix poplar and willow in the same plantation — although I have not discovered whether they are thinking of alternate rows, blocks or intimate mixtures. It may be necessary, in the long run, to split up energy plantations around the farm, interspersed with other crops and grassland.

A great deal of information about rust is contained in ETSU report number B1258 by the specialists at Long Ashton, called 'Evaluation of the biology and importance of diseases and pests in willow energy plantations'. The report contains a large amount of detail and observations from experiments, and the recommendations at the back are encouraging for prospective growers of short-rotation willow coppice. At present, there is no evidence to suggest that anything other than rust and occasional insect pests are real threats to the crop. However, growers should be aware that the importance of rust and insect attack can vary between seasons and locations. There is no guarantee against one organism or another becoming a general problem in the future: a keen eye must be kept — especially when anything new appears to be causing damage. In the last two years, for example, there has been growing concern about localised but damaging attacks by the brassy beetle in established willow. *Melampsora*, says the report, has to be regarded as a major threat to a developing willow energy industry. Even so, rust should not be seen as a reason to discourage a considerable expansion of willow energy cropping in the UK. Great advances in knowledge of the disease already suggest possibilities for practicable and economic strategies for rust management. The Long Ashton specialists go on to give some advice: avoid planting large blocks of one clone, for it is much better to plant a stand of mixed clones, or even mixed species. It is sound scientific practice, even if various mixtures have not yet been thoroughly tested.

Severe rust epidemics can arise from very small origins — given a susceptible crop and a favourable year. The worst threat comes from early season infection, and the proximity of larch may lead to an increased risk of infection, because it is one of the shared hosts of rust.

References

1   D. V. Clay and R. I. Parfitt, 'Evaluation of residual and foliar-acting herbicides on poplar and willow cultivars', ETSU report, B/N2/00132/REP, 1994.

2   S. Ledin, Proceedings from conference on 'Short-rotation coppice — growing for profit', RASE/WEDG, 24 March 1993.

3   M.-L. Tardif, 'Overview of short-rotation forestry in Canada', IEA report from ETSU, workshop and study tour, Sweden, 2–4 March 1993.

4   David Clay 'Weed control and soil management systems for rotation coppice: present knowledge and future requirements', ETSU report, B/W5/ 00211/REP, 1993.

5   David Clay, Proceedings from 'Short-rotation coppice — growing for profit', RASE/WEDG, 24 March 1993.

6   Banks/SWEB, Growers' meeting, 10 January 1994, The Bull, Long Melford, Sudbury, Suffolk.

7   E. Stenhouse, Conference, 'Wood, a new business opportunity', Cambridge, ETSU, 1993.

8   Personal communication, 1994.

9   'Talking arable', *Arable Farming*, May 1994.

10   See note 2.

11 J. Jobling, 'Poplars for wood production and amenity', Forestry Commission Bulletin no. 92, 1990.

12 D. J. Royle, T. Hunter and M. H. Pei, 'Evaluation of the biology and importance of diseases and pests in willow energy plantations', ETSU report, B1258, 1992.

CHAPTER 8

Cheap to Feed

Compared with cereals, grass, vegetables or forest trees, short-rotation coppice is — as in so many of its requirements — unusual when it comes to fertiliser. Apart from water, for which the crop has a requirement similar to that of cereal and root crops, short-rotation coppice has a less demanding appetite for other nutrients from the soil than food crops. To give just one example, it needs only about one-fifth of the annual requirement of cereals. A further important advantage is the possible use of the crop, with its fibrous root structure, to absorb nutrients from slurry, sewage and other sludges, acting as a biofilter while benefiting from their fertiliser value.[1]

Soil testing will indicate where fertility is low, and there is plenty of experimental work to demonstrate that coppice will benefit from fertilisers where this is the case. In general, however, the crop requires little by way of fertiliser and, indeed, the application of nitrogen sometimes actually reduces yields because it stimulates more competition from weeds. Because short-rotation coppice is harvested after leaf fall, most of the nutrients within the system are recycled — leaf litter can contribute over 130 kg per hectare of nitrogen.[2] A low value energy crop cannot afford much input of artificial fertilisers and, since harvested wood is primarily carbon, which is derived from the atmosphere, it doesn't need them. The small amount of potassium removed can be returned as wood ash from the processed woodchips.

There seems to be some doubt as to how much nitrogen is removed at harvest. It has been estimated to be only about 30 kg per hectare by some research workers, although others have

110

suggested nitrogen removal at up to 135 kg per hectare per year and phosphorus, potassium, magnesium and calcium at 16, 85, 12 and 208 kg per hectare per year respectively. Nutrient removal rates in trials done by the Forestry Commission were up to 135 kg per hectare per annum of nitrogen and 16 kg of phosphate. At these rates, says Paul Tabbush of the Commission's research division, nutrients are unlikely to limit growth on fertile sites, at least for the first few cutting cycles.[3]

Work at Aberdeen University suggests that if fertilisers are needed at all it will be nitrogen that is most vital during the life of a short-rotation coppice plantation. Once the canopy has closed, says a report for ETSU, a large proportion of the nutrient requirement is met by recycling leaf matter and fine root turnover. It recommends that 3–4 kg of nitrogen should be applied for every oven-dry tonne of biomass removed from the site. The time to apply fertilisers is directly after harvest, preferably before the shoots grow over 30–50 cm tall, when there is a high nutrient demand. Slow-release fertilisers are preferable to reduce nitrate run-off.[4]

Stig Ledin and others at the Swedish University of Agricultural Sciences at Uppsala have found that an average application of 60–80 kg of nitrogen, 10 kg of phosphorus and 35 kg of potassium is suitable. The principle, says Stig, is that you should replace the nutrients that are taken away in the stem biomass (*note*: that does not include the leaves, which fall back to the soil). In what Stig Ledin calls the 'temperate climate of Sweden' he estimates that about 1 per cent of soil nitrogen is mineralised every year. In the year of planting they advise against using little or any added fertiliser because it may help the weeds more than the *Salix*. However, in the early years of the crop, before leaf litter has begun to build up, a little extra fertiliser may be needed — depending on soil type. This is better applied in two doses, rather than giving the weeds a feast and also risking nitrogen being leached. Spreading fertiliser on such a lofty crop may be difficult, once it has become established. 'In the year after planting and after each harvest, applications can be made using conventional equipment,' say the Swedes, but in tall stands they found it necessary to use spreaders which could operate above the crop. Alternatively, smaller machines which could sneak along between the rows might be used. Both techniques are practised in Sweden.[5]

A great deal of detail about fertilisation can be found in Stig Ledin and Agnetha Alriksson's contribution to a handbook on how to grow short-rotation forests produced at Uppsala. This loose-leaf folder relates experiences from seven different countries on both sides of the Atlantic and is full of gems.

THE GLORIES OF SLUDGE

What a superb partnership — a profitable mainstream crop whose favourite fertiliser is something that nearly everyone is trying to get rid of! Short-rotation coppice thrives on farmyard slurry, sewage sludge and other sludges — when it is needed. The quality of sewage sludges is improving, and as water companies pursue polluters who still allow heavy metals to leave their premises by way of watercourses, this quality will improve even more.

Sewage sludge is also carrying with it a great deal of government finance at present. For example, ADAS will lead a £1.7 million research programme into the beneficial use of sewage sludge on farmland over the next four years. (The reason for this is that the UK has agreed not to dump any more sewage sludge into the sea from 1998 onwards.) The work will be based at Rothamsted Research Station and the Water Research Centre, with six sites across the country to study the effects of the sludge on soils and the implications for productivity and long-term soil fertility.[6]

The point is made by short-rotation coppice specialists that sewage sludge should only be applied if it will increase tree growth. The 'dumping' of sludge should be avoided, with only sites where the soil nutrient status is low being considered. Digested sludge is preferable because it doesn't smell so unpleasant and is less likely to contain pathogens. Cake sludge can only be used when it can be incorporated before planting, but liquid sludges can be used both before planting and after harvest. Willow is reasonably tolerant of comparatively high soil metal levels and responds well to sludge application.[7]

Three years ago 465,000 tonnes of dried sewage solids were spread over some 56,000 hectares in this country. By the year 2006 the aim is to dispose of nearly one million tonnes on the land. Some of this will be free although some water companies, like Wessex Water, have begun processing it into pasteurised nose-friendly granules with 92 per cent dry matter, for which you will be expected to pay something like £10 a tonne, including spreading, on arable land. Whether it will prove possible to spread granules on newly harvested coppice land remains to be seen, but it should not be beyond the wit of woman or man. These granules provide slow-release nitrogen, with only a fifth of their 3.5 per cent nitrogen emerging in the first year.[8]

Everyone agrees that to get the highest yields from short-rotation wood crops you need fertiliser. Inorganic fertiliser, because it requires high energy input during manufacture would seem to defeat the purpose of growing an energy-efficient crop. Far better to use organically-based nutrients such as sewage sludge, farmyard slurry and manure, poultry manure, silage effluent, putrescible town refuse and food processing wastes — they all contain what coppice needs.

The Department of Trade and Industry (through ETSU), and the Anglian and Severn Trent Water Service companies have been funding studies (due to continue until 1997) into how these organic wastes can best be used on coppice crops.[9] They wanted to ensure that three objectives are maintained: economic benefit to the crop; sound environmental practice; and an economic and convenient destination for the waste products. It is very important that plant nutrients do not find their way into watercourses and short-rotation coppice is one of the most efficient barriers to leaching. It is to be used in buffer zones to protect rivers in several of Britain's major river valleys.

Short-rotation coppice may, therefore, be a very attractive crop for water companies to grow on their own land, or in co-operation with farmers and landowners: a symbiotic business relationship in which farmers convert waste into energy with a little help from the sun and short-rotation coppice. You cannot, however, just go sloshing sludge around haphazardly. Controls exist in the form of the 'Sludge (Use in Agriculture) Regulations 1989', and the 'Collection and Disposal of Waste Regulations' of 1988. Sewage sludge definitely works on trees, however. On trials carried out in Wales by British Coal, Welsh Water and the Forestry Commission the growth rate has been tripled on areas of reclaimed land. In this case sludge was injected into the ground before the trees were planted, and this could be a possibility for short-rotation coppice, too.[10]

Mark Aitken of the Scottish Agricultural College puts a value of some £30 million a year on the UK's annual output of sewage sludge. Scotland alone produces three million wet tonnes of it (which could dry down to 100,000 tonnes), of which two-thirds at present is dumped at sea. Local authorities will have a headache in 1998 when this can no longer be done. Strathclyde, for example, will have an extra 2.5 million tonnes to dispose of. Some will end up in landfill sites, but most will have to be recycled as fertiliser. Mark Aitken reckons that an application of 50 cubic metres per hectare could be worth £33 per hectare, supplying 44 kg of nitrogen, 65 kg of phosphate and 3 kg of potash. He warns that there could be a build-up of heavy metals in the topsoil, which could pose a risk to sheep.[11] These metals are not a threat to coppice, as far as we know, although they must be taken into close consideration when sewage sludge is being regularly applied to a piece of land. We do know that even willow will be damaged by very high levels of metals and the danger is that these could be concentrated in a plantation regularly dressed with sewage sludge.

THERE MAY BE DANGER?

Very little of the metal is taken up by the stems of the crop. Some cadmium and zinc is absorbed and remains in the ash, after the crop has been burned to release energy. If the ash is used as fertiliser, back on the plantation, it could exacerbate the contamination. There is also a possibility that the ash might be condemned to disposal as toxic waste — although this is un- likely, because levels of metals in sewage sludge have decreased very markedly over the past 30 years. Scientists are divided as to their current importance, so it is best to be on the safe side and make sure that any sludge used as fertiliser on short- rotation coppice is properly analysed before being spread. This is standard practice by the water companies.

But let us put this hazard into perspective. At the 1994 English Royal Show, in the demonstration area for biomass production (don't miss it at the next Royal Show) Drusilla Riddell-Black of the Water Research Centre told me this: 'Out of the three parts of a willow tree — the roots, stem and leaves — the stem accumulates the lowest concentration of all the metals you may find in sewage sludge. Willows take up cadmium and a little zinc from the soil — but to demonstrate just how little, let us use this example. In a "hot" site (i.e. heavily contaminated with heavy metals) it would take 150 years of harvesting willow to reduce the soil concentration by 50 per cent. For zinc, it would take 7,000 years!'

References

1 ETSU, 'An assessment of renewable energy for the UK', HMSO, 1994.

2 M. Dawson, 'Production and utilisation of biomass from short-rotation coppice in Northern Ireland 1974–1988'. Dept of Agriculture for Northern Ireland, Loughgall, Co. Armagh, 1988.

3 'Coppiced trees as energy crops', ETSU contract, E/5A/1291/2285, 1993.

4 C. P. Mitchell, J. B. Ford-Robertson and M. P. Watters, 'Establishment and monitoring of large-scale trials of short-rotation coppice for energy', ETSU report B1255, 1993.

5 'The development of short-rotation forestry in Sweden', conference at NAC, 24 March 1993.

6 'Benefits of sewage sludge to be investigated', *Arable Farming* magazine, May 1994.

7 G. Hunt, 'Energy forestry in the forest of Mercia', ETSU B/W5/00241/REP, 1993.

8 E. Penny, 'A nutrient not to be sniffed at', *Crops* magazine, 16 April 1994.

9 D. Riddell-Black, 'Short-rotation energy coppice production using sewage sludge', proceedings of 'Wood — energy and the environment' conference held at Harrogate, 9–10 September 1992.

10 R. Bedlow, 'Trees grow faster on sewage diet', *Daily Telegraph*, 3 January 1994.

11 '£30m handout of free fertiliser from local authorities', *Farmers Weekly*, 11 March 1994.

CHAPTER 9

Harvest: to Chip or not to Chip?

Harvesting of short-rotation coppice is done in the winter, after the leaves have fallen. It can be a muddy job but the network of roots strengthens the soil and the season is comparatively long.

The cost of harvesting can amount to as much as 70 per cent of the delivered fuel costs[1] and there's still a lot of work to be done to bring these costs down, but prospects for this are good. Already, costs have come down from £27 per oven-dry tonne to £11. Those in the know expect the cost of harvesting to be somewhere around £150 per hectare, if the cut-and-chip method is used with a forage harvester, regardless of yields: as yields increase, so harvesting costs will decrease too.

It is most likely that harvesting of short-rotation coppice for fuel will be done by an agricultural contractor — especially one with grassland connections — because self-propelled forage harvesters can do the job without much adaptation; and since the harvest takes place from December to February, it will not clash with the silage harvest. As the area under short-rotation coppice builds up — and this could happen very fast when the NFFO contracts enable electricity generating companies to enter into 15-year contracts with farmers — woodchip merchants like Banks of Sandy say they will be organising harvesting of short-rotation coppice (using Claas machines) for farmers who don't have suitable equipment.

Fitted with a German-designed header, forage harvesters like the Claas Jaguar can cut and chop short-rotation coppice surprisingly easily, even when the coppice has been growing for three or four years. The header fits the Claas Jaguar 860 and 880 foragers well because of their wider and stronger chopping drum and higher power output.

The crop is pushed forward by a frame above high speed rotary, hydraulically powered, cutting saws which sever the rods close to the ground. (It's essential to cut the stool cleanly so that damage to the plant by frost and disease is avoided.) The crop then goes, butt end first, to feed rollers in the forage harvester. The way the machine is set up allows even the tallest of short-rotation coppice crops to be harvested, and with twin-row planting, as practised in Sweden (and which is beginning to be adopted here), Claas expects that the modified machine will be able to harvest 400 hectares per winter.[2] The chips are then blown into forage trailers with high sides and towed away to storage by tractors.[3]

Farm contractors may find it worthwhile to invest in specialised machinery such as the Swedish Austoft harvester, which is a massive sugar-cane harvester adapted for coppice. It is self-propelled and runs on tracks. I saw it working at Robert Goodwin's Ashman Farm at Kelvedon in Essex early in 1994

The massive Austoft harvester at work on Robert Goodwin's farm at Kelvedon in Essex. (Robert Goodwin)

Short-rotation coppice growers, attending an ETSU/Forestry Commission harvesting demonstration at Ashman Farm, Kelvedon, were
relieved to note that despite the wet weather the Austoft did not sink
into the mud but harvested very capably. (Robert Goodwin)

during a very wet winter. I remember John Seed of Border
Biofuels heaving a sigh of relief when the Austoft effortlessly
started work, despite the very muddy and wet conditions. He
said it had been a real worry whether it would be possible to
harvest coppice in January and February in the UK. The
demonstration, organised by the Forestry Commission on
behalf of ETSU, reassured many of the growers and enthusiasts
for short-rotation coppice.

The root system of both willow and poplar at Ashman Farm
seemed to create a raft on which even this very large machine
and the tractors and trailers shadowing it stayed afloat very
happily. However, the headlands did get somewhat chewed up
and this emphasised the need to plan the plantation layout
so that headlands and rides are on the firmest parts of the
field. The point is made by the Swedes that heavy harvesting
machinery will cause compaction even if it is on tracks; they try

to harvest when the soil is frozen solid. They do not consider the root mat capable of protecting the soil against deep compaction and recommend keeping tractors and trailers off the field as much as possible. The Austoft is reputed to cost something like £250,000 and can munch its way through very large areas of coppice in no time, so each machine would need to be based at a centre related to a woodchip-fuelled power generating plant.

The Fröbbesta rod harvester at work in Sweden.

For smaller operations such as local heating projects the other machine being demonstrated that day was the Fröbbesta rod harvester. It is pulled by a medium-sized tractor and cuts two rows at a time. Small circular saws cut the coppice rods off neatly, close to the base of the stool, and the rods are held upright as they are moved back towards the platform at the rear of the machine. There, they fall over flat and build up until the platform is full. At this juncture an hydraulic ram pushes the rods off on to the ground where they can be gathered up. The Fröbbesta is a cheaper machine with a much lower work rate: it makes small bundles of rods, although a new version will have a larger load platform able to make bigger bundles, reports Damian Culshaw of ETSU, after a visit to Sweden.

Picking up rods at the field side. (Damian Culshaw, ETSU)

But what can be done with the rods that the Fröbbesta has cut? They can be stacked on the headland to await a chipper. Chipping can be delayed until the sap has dried out but that makes the rods much tougher to chip. The grower therefore has a dilemma, and it is something which the UK short-rotation coppice industry has not yet worked out satisfactorily. What it needs is practical farmers to come up with solutions, which they invariably do once a crop begins to become better established.

For the present, forage harvesting and chipping seem to be the easiest option although, as we shall see in Chapter 10, this creates difficulties with storage and drying of the wood-chips because the moisture content needs to fall from the 50–60 per cent moisture level at harvesting to 20–25 per cent. Dr Paul Maryan of ETSU reported to the Cambridge conference in October 1993 ('Wood, a new business opportunity') that it was in the harvesting operation of short-rotation coppice that the greatest reductions in cost had been made. He said that originally it was thought that a bespoke coppice harvester would be needed — and the Loughry prototype was a result of that belief — but the Claas forage harvester had offered a less expensive alternative, being cheaper by some 75 per cent. He said that Claas were developing a special header for the machine, to enable it to harvest twin rows. Even the unmodified

The Claas forage harvester gobbling up a light crop of short-rotation coppice. Using such a forager has slashed the cost of harvesting. (ETSU)

forage harvester was capable of accepting stems with a butt diameter of 50 mm. This would mean that poplar would need to be harvested on a two-year cycle and willow on a three-year cycle, to avoid the stems getting too fat for forage harvesting.[4] There is some doubt as to whether poplar could withstand the stress of being harvested on such a short rotation.

IT'S THE BENDER

Stacking the rods at the side of the field does reduce moisture fairly rapidly but the wood hardens up, making chipping more expensive. But there are new ways of harvesting which may alter the whole picture. A Swedish machine, with which Murray Carter, the Yorkshire farmer and coppice cutting supplier, will be experimenting in this country cuts the rods at the base and then bends them in half as they are pulled into a round chamber. This produces a continuous sausage of compressed rods which goes into the chipper that forms part of the machine. The Bender harvester will cut and chop some 16 dry tonnes an hour, on average: this would give a harvest rate of around 0.4 ha an hour (one acre per hour). Maximum output is 20 dt/hr. The machine needs a 100 kW tractor and fuel consumption is claimed to be 15–20 litres per hour. The Bender fits on a front-mounted three-point linkage. So far it has not been tried for poplar in the United Kingdom, although it was due to be tested in the winter of 1994/95. The continuous sausage could be cut into standard lengths and tied into bales, after which the material would be handled mechanically, using conventional forestry equipment.

The Bender harvester is built by Messrs Wilstrand, in Sweden, who developed the step planter (see Chapter 6). Another harvester, developed at Loughry College in Northern Ireland since 1979, ties the rods into bundles (which are ideal for the beanstick market). In its present form the Loughry harvester is very expensive, but a commercial model could be developed. The question is, though, is chipping really necessary? Could the bales or bundles of rods not be treated in the same way as the bales of straw which, in Sweden and Denmark, are burned whole in district heating systems? The bales are handled automatically by computer-controlled gantry cranes and fed

The shape of things to come? The Bender harvester at work in Sweden. (Stig Ledin)

into boilers, on rails. Well, apparently this does not work for coppice rods. To get the optimal amount of energy from them they must be chipped, because a maximum surface area of the biomass must be exposed during the combustion or gasification process to produce high enough temperatures.

Should the rods be chipped in the field, then, or at the generating station or heating plant? I have no doubt that the solution will become apparent as engineers and farmers put their minds to it. One could visualise, for the smaller operation, a farm contractor owning a Fröbbesta or Loughry harvester *and* a chipper, and doing the rounds of local farms twice: once to harvest and once to chip. Although foresters tend to scoff at pto-driven chippers used by farmers (foresters generally use specialised and much larger machines), there are many chippers on the market. At Drayton Park in Northamptonshire, where Drayton House is heated with woodchips, such a machine has proved very useful, according to John Lockhart, the land agent who helped set up and run the system.

While waiting for your crop of short-rotation coppice to grow you may have to find an alternative supply of woodchips to cover the first four years. Thinnings, loppings and toppings from local woodland may be the answer, and a small chipper on the back of a tractor could be very useful. It will, of course, be much more expensive than having a contractor in with his modified forage harvester, hoovering up the short-rotation coppice, but it will leave local woodland in a much healthier condition and add substantially to its value.

COMMINUTION — NOTHING TO DO WITH STALIN

Another possibility might be the use of a grapple-fed tub-grinder for chopping up the rods, bales or bundles. This chopping up is called 'comminution' (another term which will become a common part of farming parlance in the next year or two). Boilers and gasifiers can be fussy when it comes to 'woodchips'. They do not take kindly to a mixture of sawdust, 5 cm fingers of willow rod and great chunks of log. What is needed is a standardised product from the farm.

There are no agreed standards as yet, although the industry's new umbrella body, British Biogen (set up by the Wood Energy Development Group, the NFU, the CLA, SNFU, TGA, AIEP, electricity generators and other interested parties) has as one of its targets the establishment of such standards. ETSU commissioned engineers FEC Consultants Ltd to produce a report on the subject. Called 'Wood fuel standards' (ETSU report B/W3/00161/REP) and published in 1994, it provides a starting point for industry decisions. Engineers and farmers will have to work closely together to design, build and use equipment which will produce woodchips of a suitable standard and boilers, with their hoppers and automatic-feed systems, to cope with the variation expected within those standards.

Perhaps it is best to examine what is already being done in Scandinavia as a likely indicator as to what may happen here, come coppice harvest time. Stig Ledin of the Swedish University of Agricultural Sciences told delegates to the NAC conference on short-rotation coppice in March 1993 that some large district heating plants in Sweden were able to burn undried coppice with a moisture content of around 50 per cent. In these cases direct chipping and transport to the factory were the methods employed. If the burning is to be done on the farm (for domestic heat, or heat and power), then drier chips will be needed. Rods

Instead of the annual visit by the threshing machine, short-rotation coppice farmers may need to hire a very large industrial grinder such as this Gannon HD 10 to produce woodchips. (Derek Overton, Gannon UK Ltd)

'Very useful' tractor-mounted woodchipper at Lionel Hill's farm.

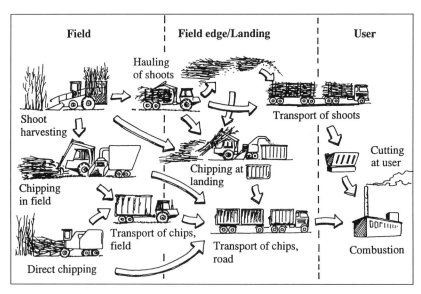

Choices of harvesting and handling systems for short-rotation coppice as practised in Sweden. (Sigge Falk)

that are harvested whole are allowed to dry in piles for one summer to bring moisture down to 30 per cent by the next winter.

HARVESTING AS PLANTING MATERIAL

When the crop is being harvested for planting material, there is a major expense involved in keeping each type of clone separate. Several Swedish machines are available for this process, including one made by Stegerslått which is mounted on a tractor three-point linkage; another is a trailed machine sold by Fröbbesta. In Sweden they have found that two skilled people, cutting selected rods by hand, can often compete with machine harvesting of cuttings, and yield about 90 per cent of usable material.

When the separate clones have been harvested they are sized and cut up in the farm buildings, then marked and packed before being kept in a refrigerated store. If one cutting is dropped from the workbench on to the floor, during this process, at Murray Carter's Harrogate farm, it is not sold as named planting material because the risk of mixing up different clones is so high. I mention this to indicate that the preparation of material for planting is not something to be undertaken lightly. You are also likely to need cool/refrigerated storage for cuttings.

HOW MUCH CAN YOU EXPECT?

Yields of short-rotation coppice are increasing as planting material and the techniques of planting, establishing and growing the crop on improve. However, productivity is a result of many interrelated factors such as planting density, the period of growth which elapses between harvests, the climate, availability of water and nutrients, the type of clone planted and the effects of pests and diseases. It is difficult, then, to be specific about 'how much you should expect', in the current state of knowledge. There are numerous reports in the literature of maximum yields of up to 35 oven-dried tonnes per hectare per year — but they refer to small trial plots in southern Sweden, with optimal nutrition and irrigation, or an even more favourable situation in New Zealand. And the people responsible for growing those crops are confident they can improve on even these very high figures.

Much of the data available for the UK refers to yield after only one year of growth, when the crop is still not fully productive. Annual harvesting may not reflect three-year growth potential and, even then, full potential is not attained by the first harvest at three years — it could well increase further as the plant becomes better established. Most of the trials done in the UK have been from wetter, cooler areas; results from the drier east may differ. In 1990 a trial at Long Ashton, near Bristol, compared 25 willow and 7 poplar clones at a planted density of 40,000 per hectare on a silty clay soil. Yields from the best one-year-old stems on three-year-old stools averaged 18 oven-dried tonnes a year, from small plots.

The highest yields were from *Salix × mollissima undulata* hybrids, but yields of 10–12 odt/ha/yr were recorded for other clones. A similar trial at the same density on the peaty soils of the Somerset Levels gave yields from one-year-old rods on two-year-old stools of up to 12.4 odt/ha/yr, while on a light loamy soil in the Midlands yields of one-year-old rods on two-year-old stools were up to 12 odt/ha/yr. This gives some idea of the variation to be expected. With new clones coming through all the time, and more knowledge about how the crop should be grown, yields are projected to rise. All this demonstrates, however, the value of putting in a small trial plot to learn how the crop reacts to local conditions (and your management!).

The Somerset trial, when harvested as a three-year-old crop, gave yields on some individual plots of 20 odt/ha/yr, according to Ken Stott of Long Ashton Forestry Consultants. This figure is by no means out of reach for commercial UK short-rotation coppice production 'within the foreseeable future', according to many leading scientists and producers here.

References

1   G. Hunt, 'Energy forestry in the forest of Mercia', ETSU, report B/W5/00241/REP, 1993.
2   'Foragers convert for coppicing', *Timber Grower* magazine, summer edition, 1994.
3   Damian Culshaw, 'Status of short-rotation forestry mechanisation worldwide: workshop and study tour', IEA Report, 2–4 March 1993.
4   Paul Maryan, 'The production of wood fuel', at the conference, 'Wood, a new business opportunity', Cambridge, October 1993, proceedings published by ETSU.

CHAPTER 10

Storage, Drying and Utilisation

Short-rotation coppice is harvested between the end of November and the end of February, but the market will need it to be supplied all year round, so large quantities of a comparatively low value farm product will have to be stored for many months. Also, at harvest time, the woodchips are wet, at about 50 per cent moisture. Although in large district heating plants in Sweden they are sometimes used in their wet condition, most UK combustion plants will require drier material, with a moisture content of around 25–30 per cent.

It would be sensible to assume that storage will cost money. The question is, how much? Paul Mitchell of Aberdeen University's Forestry Department says that storing woodchips outdoors, uncovered, can cost from £3.82 to £6.01 per green tonne. Storing them under cover would cost between £5.38 and £5.51 per green tonne.[1]

Much will depend on the facilities available on a farm. Some farms have surplus barns and livestock housing in which woodchips could be stored. Some short-rotation coppice growers may wish to harvest the biomass as rods and chip when the moisture content is lower. They would then be able to store the rods outdoors. Some farmers are already making money with their buildings, for caravan storage or office accommodation, for example; others will not wish to spend the extra that it costs on chipping (i.e. comminuting) dry rods.

The experts on chipping dry rods are foresters and forest contractors. They have very large self-propelled chippers which could visit a farm and chip, rather in the way threshing machine contractors used to travel from farm to farm. Barry Hudson of the Forestry Contractors Association (FCA) says that for a small-scale operation a tractor-mounted chipper can be bought from between £2000 and £12,000. Hand fed, this will cope with 1.5 tonnes of biomass per hour and such equipment might be sufficient for a project involving some 600 tonnes per year. That quantity of woodchips would be enough to heat a large school with sports centre and swimming pool attached, and could be produced from less than 40 hectares of good arable land. But he estimates the cost of comminuting with that method at £10–£12 a tonne.

Barry Hudson prefers the 'tub-grinder' option, as long as the tub-grinder is strong enough to cope with the larger dry rods. This machine grinds the wood down to 'hog fuel' size, which is easy to handle in an automatic-feed system in a heating plant. I saw one such tub-grinder at the Royal Show in July 1994. It was the Haybuster machine, from the USA, and prices start at £35,000 for a tub-grinder fitted with a 110 hp John Deere engine, from Gannon UK Ltd at Welbourn, Lincoln. At that price you can see why it's a contractor's tool — unless you have a vast area of coppice and woodland. I understand that this machine can also grind up domestic waste for burning as biomass — but that's another story!

'Comminution should not cost more than £5 a tonne — and to achieve that you really need a throughput of 10 tonnes per hour,' Barry Hudson told a seminar held at ETSU in the spring of 1994. It may be that if you have the 600 tonne-sized project you can fit the woodchipping exercise into a period when you would otherwise have little to do and can carry the cost. Every producer will have to resolve this, in his or her own way. It may be cheaper to get the local forest contractor to come and do the job for you — as long as your budget can cope with the extra cost. You will find the FCA's address at the back of this book.

Stacked wet woodchips heat up and lose value for the first few months, perhaps at a rate of 5 per cent per month during this period, although Swedish experience shows that the moisture content will have fallen to around 30 per cent by the following winter.[2, 3] Kept in piles outdoors they will get wet again and lose more energy, so proper storage is essential to minimise spontaneous combustion and further microbial

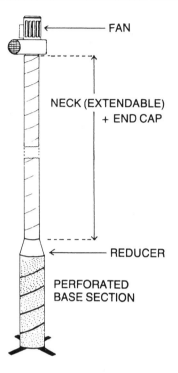

FAN

NECK (EXTENDABLE)
+ END CAP

REDUCER

PERFORATED
BASE SECTION

A pedestal dryer
that might be suitable
for drying woodchips.
(Martin Lishman)

Pile-Dry Pedestals from Martin Lishman. These are in use in a 1000 tonne grain store. The fan is moved from one pedestal to another as cooling/drying takes place. (Martin Lishman)

degradation. Covered storage actually costs less than uncovered storage in the long run, as the gain in energy offsets the provision of covered storage.[4]

One of the few people to have experience of this problem in the UK is John Lockhart, of chartered surveyors Samuel Rose. In his work at Drayton Park in Northamptonshire, where the mansion is heated with woodchips from the estate, John tells me that they were fortunate in having old Dutch barns in which to store woodchips. 'But if you haven't got that just about any sort of roof cover will do,' he says. 'A plastic tunnel (with open sides if possible) or a marquee-style building. Netlon mesh cages will allow piles 3–4 metres deep of woodchips.' The outward pressure from a pile of woodchips is apparently not high at that depth.

'We have found that if you harvest little and often over a period, the chips dry quickly and there's very little heating up in the stack,' he says, 'and dry matter losses are very small.' It is the heating up which causes the loss. 'If you are harvesting a lot at one time, then you'll need to blow low volumes of air through the chips, from the bottom.' John Lockhart says that there are various ways of keeping a pile of woodchips cool in a building. If the chips are against a wall, for example, a drainage pipe can be laid along the bottom of the wall to allow the passage of air. A tractor bucket is a useful tool to turn piles of woodchips over to allow them to cool off. For John's own purposes, he uses a figure of £0.80 per cubic metre as a storage cost.

STORAGE SOLUTIONS

This whole problem of storage is one which farmers may solve faster than the research engineers. Already there are many different ways of blowing air through bulky piles of farm produce — for example, a grain drying floor with air ducts; or air ducts which can be laid on a barn floor; or, indeed, a large diameter drainage pipe put in place before the crop is brought in, with a low-powered extractor fan on the end. It may well be that existing facilities could be used or adapted for both drying and storage, without any extra cost to the farm. This newly fledged home-grown energy industry does realise, however, that for most producers, storage could be a problem and research is going on at Silsoe Research Institute. Martin Nellist has a four-year programme of research there, financed by the Department of Trade and Industry (which also deals with energy).

Martin Nellist says that by investigating the science of how woodchips dry, he hopes to be able to come up with firm advice on general principles for drying them. 'Wood fuel is 15 times the bulk of diesel oil,' he told us at a seminar at Harwell in April 1994. 'Coal is only 1.7 times bulkier than diesel.' This made wood fuel far more expensive to move around and store. Wetness was also a very important factor. 'Wetness reduces value. To dry it, we need to bring moisture down from 50 per cent to about 25 per cent. This represents a very large quantity of water and it needs a *lot* of energy to do this: but you can't really afford to buy very much energy for this purpose in such a low-value crop.' Martin said that to evaporate moisture in a combustor would need 2.45 gigajoules per tonne of water removed: that is the latent heat which is needed before water evaporates. To evaporate moisture by conventional drying would require not less than 3.3 gigajoules per tonne of water evaporated.

Martin Nellist looked at some of the alternatives, such as getting nature to dry the crop for you. 'You can harvest and store as whole sticks; or you can harvest and chip and then burn immediately; or dry it using waste heat from a generation plant. You can store chips in bulk — allowing them to dry, ventilating with cold air just to keep them cool: this will not only reduce losses but dry them (slightly) at the same time.'

He agreed that drying unchipped rods outdoors is cheap but means double handling; and it was difficult to handle and chip. However, 'if you harvest and chip in one operation, and burn immediately, there is no capital cost for storage but you waste a lot of energy evaporating off all that water.' It was occasionally possible to do it in this way without serious economic loss, if there was a use for low grade heat, such as for domestic heating, and if a sophisticated heat recovery system were used.

HEALTH DANGER

Allowing piles of woodchips to heat up to above 35 degrees Celsius was dangerous, said Martin Nellist, because it allowed the development of fungi which produced the spores causing the disease 'farmer's lung'. Higher temperatures also meant higher losses. Done carefully, though, the heating-up process can be used to help the drying. 'Allow the chips to heat but as soon as the temperature reaches 35 degrees blow air through them.' But this would lose dry matter, he warned, and it was risky. One farmer at the seminar said that he had 100 tonnes of woodchips piled on top of ducts, on the floor. At first he'd used a diesel blower several times a week, and great clouds of steam would come out of the pile, but after three months heating slowed down a lot and he was able to cut the use of the blower to once a week.

PUBLIC PERCEPTION OF WOODCHIPS ON THE MOVE

It is not economical to move woodchips over long distances in the way gas, oil and coal are moved. The nearer short-rotation coppice is grown to where it will be used, the better. If farmers' tractors and trailers can deliver, so much the better, because it is likely that most demand will be in the depth of winter when trailers and big tractors generally have less work to do. For local heating markets this may not prove to be a problem but Paul Mitchell, at the ETSU seminar, gave some examples of the quantities needed for power generation, illustrating just how much biomass would be needed.

'One megawatt needs 7500 tonnes of oven-dried woodchips per year. That is 20.54 tonnes per day and that represents 123.24 cubic metres of bulk. So a 10 megawatt generation plant will need 1232.42 cubic metres of woodchips per day! That's 20 lorries and probably 7 woodchipping machines working every day.' Such quantities could lead to problems with the planning authorities and local pressure groups who might resist this kind of increase in traffic. There is also a storage problem, said Paul Mitchell. 'If you are going to store one week's supply of woodchips for this 10 megawatt plant, the 8626.8 cubic metres will make a pile of chips 50 metres long by 50 metres wide by 4 metres deep! When you have large-scale operations you have strategic problems — so before going too far ahead, just visualise how the full picture will look on the ground — the drying, the storage and the transport.' Transporting woodchips is bad for business, because woodchips are of inherently low value. John Lockhart was charging himself £3.30 per dry tonne for haulage within a 20-miles radius in 1994, based on current prices at the time.

In defence of transporting woodchips on the roads, Paul Mitchell said that the Timber Growers Organisation maintains that the use of public roads for timber is *not* unusual or special. It has always been done. Already in the West Country, however, local authorities have begun restricting the size of vehicles used for moving wood and may object if there is a move towards larger transporters. Dr Paul Maryan of ETSU, speaking at the same seminar, said that planning considerations and the public perception of the supply and utilisation of biomass were potentially more difficult obstacles to progress than the technical problems. 'We must plan sympathetically and carefully,' he said. 'At this stage the public doesn't have a view about biomass for power, but just look at the difficulties encountered by wind energy projects and take warning!' Terms like 'industrial development' (factories) and 'heavy transport' were emotive and, once put about, created a difficult atmosphere. There would be conflict about where to site generation plants, as planners tried to fit them into a suitable local environment and electricity producers attempted to site them for easy access to the national grid.

'People hate the thought of trees being cut down. There's the visual impact and the noise, too,' said Dr Maryan. 'Even though clearing and chipping leaves cleaner woodland, the ruts and damage to flora may upset local conservation groups and walkers. We need to learn to communicate with all these people — planners, customers, local residents and the general public. If people know and understand what you are trying to achieve, they are more likely to tolerate what you are doing and support your endeavours,' he said. He quoted the example of schemes for processing domestic waste for energy which have failed to get permission because people had assumed that there would be a lot of smoke.

THE COST OF WOODCHIPS

During the first few years, while the short-rotation coppice is growing, projects will need woodchips derived from local woodlands and forests. For the first time, says Major Edward Stenhouse, the biomass consultant who helped set up the five demonstration farms across southern England, woodland owners will have competition between buyers of their loppings, toppings and thinnings. However, at the Harwell seminar it was felt that landowners would receive less for their wood for biomass-energy projects, than for pulp: perhaps £3 per tonne. A much wider range of quality would be usable, especially if a chipper worked on site and contractors could leave the wood-land much improved. How much, though, would it cost to obtain the woodchips for burning?

John Lockhart of Samuel Rose, who helped set up the heating project at Drayton Park in Northamptonshire, taking part in discussions at the seminar, felt that £10 per cubic metre of chips was 'quite expensive'. But that's the charge they put on it at Drayton. 'Bigger users don't want to pay much more than £30 a tonne, at 4 cubic metres per tonne and at 25 per cent moisture. That's about £7 per cubic metre. If you can get the moisture down to 20 per cent you'll get about 4.5 cubic metres per tonne.' He said he was including the benefits of the woodland grants in full: these varied drastically, depending on where you lived. However, he said that to work out what you could afford to do, start with what power companies or other woodchip-for-energy users would pay — perhaps £1.50 or £2.00 per gigajoule (and you can say that there's roughly 19 GJ per oven-dried tonne of chips) and then work backwards from that.

AT THE PLANT

Once the woodchips have arrived at the place where they will become fuel, how will they be received by the engineer in charge? There's a parallel here with livestock going to a slaughterhouse — an area fraught with danger of disagreement between producer and processor. How do you define a wood-chip? At the Harwell seminar Fred Dumbleton, a conversion technology specialist on ETSU's staff, who was formerly with Rolls-Royce, said that definitions for wood fuel were scarce. You could, for example, obtain wood fuel as 'small roundwood', 'forest residues', 'sawmill residues', 'wood waste' or 'short-rotation coppice'. If someone offers you 'sawdust', what do they mean by it, and what would you expect it to mean? Would it include chunks of timber or slabs of bark? 'You have to discuss this with suppliers,' he said. If what was on offer did not suit your plant it would be possible to process it further, but that would be expensive.

While growers of short-rotation coppice were fighting to get their costs down, they should not ignore the requirements of the market. 'Producers cannot write the spec for woodchips on their own,' he said. 'And neither can users.' On very large projects the material being produced for fuel must suit the machinery through which it will be processed; but, on the other hand, the machinery should be designed to cope with any variations which may be expected. If there were a lot of pine needles in among the woodchips, for instance, then this should be taken into consideration when the plant is being designed. There is a need, said Fred Dumbleton, for national standards of wood fuel.

'It's about limiting variability,' he said. We need to be able to define the type of wood and its source. It would be fatuous for a farmer or landowner to say, 'My woodchips are from organic free-range willow', for example. We had to avoid 'silly specs', said Fred. Much more important would be the moisture and ash content. Once a spec had been agreed, producers must aim for variability of less than 5 per cent. 'Most wood is similar in its make-up but you must get regular quality because the user cannot afford to analyse every lorryload. Checks on quality must be simple and cheap and the user must be able to trust the producer.' Moisture content was all-important because it affected the calorific value directly. It also affects the type of combustion plant that is installed. The plant must be able to cope with the expected range of moisture contents but the producer must then stay within those limits. Ash varies little, but soil contamination of woodchips can cause problems, especially in large plants, and this needs monitoring. Thick bark easily picks up sand and soil, and while there's no particular problem with burning bark — it has good calorific value — what it carries may cause difficulties. Bark from short-rotation coppice was less of a problem because the crop was cut while still young. Fred Dumbleton gave a warning about using wood that had been treated with creosote or other preservatives because this was legally defined as 'wood waste' and combustion of this came under strict laws and regulation. He reminded delegates that with a combustion plant over a certain size, Her Majesty's Inspectorate for Pollution would have to be consulted.

BEYOND FARMING AND FORESTRY

It is unusual for farmers and foresters to take the crop right through to processing. For most of the large schemes that are being planned, in which thousands of hectares of coppice will be grown for electricity generating stations, farmers will once more be producing 'a commodity' — except where, like Border Biofuels, they own the power station and are selling electricity. However, with projects in which the farmers and landowners are adding value to their produce, they will need to enter a new field of expertise: that of engineering, boilers, turbines, generators and so on. There will be similarities to agricultural engineering, especially when it comes to feed augers, hoppers, driers and burners so it will not all be completely new.

With rural heating projects, in which farmers will be selling heat, or combined heat and power, they will be buying plant and equipment which may be new to them. Some of it is bound to be new because it has still not progressed beyond the prototype stage! An example of 'pre-commercial' plant was on display from July 1994 onwards by John Seed, of Border Biofuels: his gasifier. Using it, he says he can produce 1 kW hour of electricity from 1 kg of dry wood; and 4 kW hours of heat. The plant on display consisted of the gasifier — sturdily and simply built ('You must be able to fix it with a hammer and a wrench, for farmers', says John Seed) — and linked with it by a pipe was a diesel engine adapted for using either diesel or gas or a mixture of the two.

Gasifiers loom large in the subject of energy from short-rotation coppice. The power stations that will run on woodchips will use gasification to make combustion very efficient in large gas turbines engines. But what is gasification? Nick Barker of ETSU describes it as 'incomplete combustion'. The chamber in which the wood is burned is not supplied with enough air to burn the wood completely — only enough to convert it into a fuel gas that retains most of the heating value of the wood. This gas, an 'intermediate fuel', is then cleaned to remove tars and impurities, so that it can be burned in a diesel engine or gas turbine. Conversion efficiency from biomass to electricity, using a gasifier, is 20–25 per cent (using currently available plant, although better efficiency can be expected from forthcoming developments) — and you also have the added benefit of heat as a by-product, from the engine exhaust and jacket.[5]

GASIFIERS

Once commercial gasifiers become available they will enable small-scale use of biomass to produce both electricity and heat through semi-automatic generating sets. Nick Barker suggested that such units could be used in dairies, and by meat packers, hotels and agricultural processors. The design for this small unit cannot be scaled up dramatically but it is possible to have several units working together. It gives more reliability, because if there are six units and one fails, temporarily, the supply of electricity will not be totally lost. There are designs for much larger gasifiers, such as would be used in power stations, generating 10–20 megawatts of energy.

Another pre-commercial prototype gasifier is the one at Loughgall, in Northern Ireland, where home-grown willow is being used to heat and light a college and experience of using such equipment is being gathered. At present, I understand, the importance of having a back-up system is frequently emphasised, as staff come to grips with the vagaries of the new technology.

Gasifiers typically have a problem with cleaning up the gas before it goes into the engine which drives the generator. There can be up to 10 per cent tars in the gas, and dust too. The gas also contains carbon monoxide — as did mains gas in the old days, because it was a similar mixture of combustible gases produced from coal and coke. Cooling the gas with water also presents problems because it produces a toxic residue, making it expensive to dispose of the water. Gas from large gasification plants, such as would be used in the 10 and 20 megawatt power stations, can be used to drive turbines. Existing gas- or oil-fired boilers can also be converted to wood firing by adding a gasifier. Large-scale gasification equipment is generally more efficient in converting biomass to power, at around 70 per cent;

and with internal combustion engines operating on this gas having an efficiency of 25–30 per cent, the overall efficiency is about 20 per cent. However, the technology is still not proven. It is hoped that the NFFO scheme will encourage companies to invest in large-scale gasification plants, but at present capital costs are high, estimated at £1170 per kilowatt of energy with operation and maintenance in the order of 2p per kW hour.

In their report on 'Energy forestry in the forest of Mercia', Graham Hunt, the director and project manager of the scheme, and engineers Bernard Wilkins and Christiane De Backer, say that small-scale systems for producing energy have some advantages in rural areas. They can comprise small plantings of coppice — say, 20 hectares; a tractor-operated harvester; transport of woodchips over short distances — perhaps up to a mile or so; barn storage; and a tractor-sized engine-generator. 'At this level,' says their report, 'many of the capital and labour costs can be combined or hidden in with the other farm costs.' Other advantages of small-scale production are that small generating units, unlike the massive ones which take months to build, can be manufactured in large numbers, using mass-produced components. 'They also offer the advantages of minimal transport and marginal labour,' says the report.

The Mercia group give guidelines as to the price of such equipment, saying that direct combustion units for heat production are in the order of £100 per kilowatt for a 40 kilowatt unit and £50 per kilowatt for a 300 kilowatt unit. The problem, for the moment, is that these units cost as much as three or five times more than an equivalent gas- or oil-fuelled unit. Once the market for biomass-fuelled units increases, the price will come down and entrepreneurs like John Seed of Border Biofuels, Alick Barnes of Green Fire Energy, Bob Talbott of Talbott Boilers, Richard Parker of Nordist and groups like LRZ Bio-Energy Systems, and others, are developing gasifiers of different capacities.

Once gasification becomes commonplace it will open up all kinds of opportunities. Using gasification and internal combustion engines, electricity generation will become economically viable, according to the Mercia group. Diesel engines, they say, can achieve a higher efficiency than the most complex bespoke steam turbine but cost only a tenth of the price. They maintain that this is important because electricity is not only an energy commodity: the purchaser also pays for capacity. In other words, they pay extra for being able to have a lot more electricity on demand, if they want it. Small, low cost, wood-gas-fed engine-generators are eminently suited to meet the most expensive winter peaks of electricity demand, which occur at the period when most farmers have a little more time — indeed, at the time when short-rotation coppice is harvested. So far, though, only batch-fed units are available, which means constantly stopping and starting the process. The whole business is still (to the uninitiated) at the 'Heath Robinson' stage: enthusiasts will disagree!

Bernard Wilkins, speaking at the Royal Show in July 1994, claimed that producing electricity on farms and feeding it into the national grid could pay farmers much more than supplying large, off-farm processors. He calculated that growers could get close to the 3p per kilowatt hour production cost needed to compete with industrial electricity generators. In contrast, he said he believed that large, off-farm woodchip processors would struggle to produce power for 5p/kWhr. Countryside production would give cost savings from using existing labour, and from not having to transport chips to a different site. Bernard suggested that electricity production from small farm units could be scheduled so that generation could be run to meet peak demand. He estimated that earnings could be as high as 6p/kWhr, translating into £60 a tonne income from the woodchips. At that rate a typical farm could be growing 30 ha of biomass for three on-farm generator sets, meeting any set-aside requirement; and he quoted a typical gross margin for this system as being £450–£750 per hectare. His target cost for a 100 kW unit was £15,000.[6]

Another advocate of the small gasifier is Rod Parfitt of Long Ashton. He is, however, more cautious and has reservations about the high hopes put forward by Bernard Wilkins. 'Bernard could eventually be correct in his 3p per kilowatt hour,' says Rod. 'Swedish estimates are even lower, at 1p, but it will need volume production of cheap units to achieve this.' He feels that scheduling to produce electricity only at times of peak demand is probably a non-starter because of the need to achieve a fast return on the capital invested: this would require continued output for at least nine months of the year. 'As for 100 kW for £15,000 — or £150 per kilowatt: it's a great target but since current one-off systems cost nearer £700 per kilowatt, I can't see cost savings derived from volume production achieving an 80 per cent reduction. Even a 50 per cent reduction would be most welcome.'

Rod Parfitt is working on a gasifier at Long Ashton which is rated at 30 kW (electrical). It produces a total of 90 kW, with 30 kW as electricity and 60 kW as heat. He believes that three of these together in a unit would even out variation in gas production. 'One hundred thousand farmers in the UK could produce electricity at home and sell to the grid using these small gasifiers,' he told me. 'This would be more stable than one huge generator.' It would also mean less transport of biomass and less impact on the countryside. 'Each unit would need one tonne of dry woodchips per megawatt hour,' he said. 'One kilo of dry woodchips should give you 1 kilowatt hour: and here at Long Ashton, from 2.5 hectares of willow, giving 15 tonnes of dry woodchips per hectare per year, we should be able to produce 30 megawatt hours of electricity per year. From 10 hectares we should be able to produce 150 megawatt hours per year, sustainably. From 13 hectares of coppice we could run our 30 kW unit for 18 hours a day, every day, every year!'

Rod Parfitt's colleague, Dr John Porter (now Professor of Arable Ecology at the Royal Agricultural and Veterinary University at Agrovej in Denmark) told me that if the NFFO price for coppice electricity were to come out at, say, 9p per kilowatt hour,

Woodchipping at Voelund Co. at Holstebro in Denmark. (David Margesson)

that would be £9 per mWhr which would bring in a gross of £1080 per hectare per year. Rod Parfitt thought that these figures could be even better in the future because, at Long Ashton, on some plots of Swedish Svalöf Weibull willow clones, they had achieved yields of almost 20 dry tonnes per hectare per year. The 9p per kWhr, they emphasised, was 'just speculation'.

The very large 28 mW electricity and 67 mW heat biomass and natural gas power station at Holstebro. (David Margesson)

For further reading I recommend the report 'Energy forestry in the forest of Mercia', which is available from ETSU. It has been prepared by many of the leading figures in this new industry and it is packed with facts, ideas and solid experience. If you are an engineer and would like to know more about industrial boilers, another report is 'Forestry waste firing of industrial boilers', ETSU no. B1178, produced in 1990 by FEC Consultants Ltd of Oldham.

HEAT ALONE

For those planning 'heat only' projects — such as Home Grown Energy — Calne, with the possibility of heating a sports centre

Lionel Hill and his Talbott automatically stoked woodchip heating boilers (one is a standby) at Dunstal Court, Feckenham.

and neighbouring secondary school with woodchips from local farm woodland and short-rotation coppice — there is a much wider choice of combustion equipment because it is a much more highly developed technique. Wood-burning boilers have been used for many years, especially in joinery factories and saw-mills, producing heat from offcuts and sawdust. Automatically fed boilers with a capacity from about 150 kW up to 40,000 kW thermal output and even larger are available.

Farmer Lionel Hill at Feckenham near Redditch has one of Bob Talbott's automatically fed woodchip boilers to heat his farmhouse and swimming pool, for example; and other suppliers, like LRZ and Banks of Sandy, are also able to deliver complete systems. In this whole business of short-rotation coppice for energy, heating (and possibly providing electricity for) farmhouses and farm enterprises such as poultry, pig and dairy units should not be overlooked. If you are producing woodchips for sale to electricity companies, or to provide fuel for your own local heating project, then it would make sense to use the fuel at home, too. With fuel oil now subject to Value Added Tax, and with annual fuel bills for heat running at something like £800 to £1200 a year, it may soon be worthwhile planting an extra hectare of short-rotation coppice for home-grown home heating — as soon as automatically fed boilers have come down to sensible prices. Once the farmhouse is equipped in this way, why not the neighbours? There are many pubs, hotels, large houses, old people's homes and other large buildings with heating requirements. Farmers, foresters and landowners could easily start a 'woodchip round', filling up hoppers and empty-ing the ashbox once or twice a week locally, as required. They could also, perhaps, provide all the necessary maintenance, after training from the boiler supplier. Such a woodchip round could make all the difference as to whether a local heating project like Home Grown Energy — Calne becomes viable or not.

Border Biofuels' new gasifiers may also be used for heat production. Adrian Bowles of Border Biofuels says that a 200 kW (heat) unit should cost in the region of £14,000, including combustion unit, gas filters, boiler, augers and feed hoppers. These can be run in groups to produce higher output. The company, which will manufacture a gasifier developed on the continent, will also be offering smaller units suitable for homes, starting at 30 kW, costing something like £2000, and running up to 100 kW suitable for a farmhouse, priced in the region of £6500. A farmer could extend his or her short-rotation coppice area by one hectare to keep such a unit working, with a consequent saving of VAT and bought-in fuel — truly 'home-grown energy'. There is, however, one snag. As the law stands at present, all gasifiers, of whatever size, come under Schedule A of the Environmental Protection Act regulations. These require you to install instruments which monitor emissions; and you have to make regular reports, with the plant being regularly inspected — at extra expense. When the legislation was drafted the widespread use of pure fuel such as wood was not foreseen and it may be necessary for bodies such as British Biogen to make representations to government for the regulations to be revised. Wood combustion generally comes under Schedule B, which is much less onerous. You will need to negotiate with the local environmental health officer — whose guidance notes appar-ently do not mention willow or poplar — although Her Majesty's Inspectorate of Pollution may be able to provide you with a letter indicating that wood boilers can come under

Schedule B, and be governed by the 1956 Clean Air Act, with annual monitoring and the requirement for the boiler to be virtually smoke free.

Existing wood boilers also have certain disadvantages. One is that they run better at a constant rate. You can't easily 'turn them up' or down as you can with gas or oil. These boilers run better when the load doesn't vary more than 10 per cent or so during the year, according to Richard Parker of the Nordist Company. He says it may well be worth considering a wood boiler to take a base load capable of a 50 per cent turn-down ratio, with gas or oil providing peak loads.[7] Gasifiers may make power or heat production 'more adjustable'.

A final word on the utilisation of woodchips from short-rotation coppice. Shop around, contact two or three groups of engineers (you can find some addresses in Chapter 15) and take your time considering what you might wish to install and how much it will cost now, and in the future.

References

1   Seminar on woodchip supply at ETSU, 28 April 1994.
2   S. Ledin, 'The development of short-rotation forestry in Sweden', paper given at NAC conference 'Short-rotation coppice — growing for profit', 24 March 1993.
3   Damian Culshaw, 'Status of short-rotation forestry mechanisation worldwide: workshop and study tour', Sweden, 2–4 March 1994; International Energy Agency.
4   G. Hunt, 'Energy forestry in the forest of Mercia', ETSU report BW/00241/REP, 1993.
5   Richard Parker, 'Wood fuel combustion plant, commercial availability and operation', paper given at the 'Wood, a new business opportunity' conference held at Cambridge on 13–14 October 1993.
6   C. Abel, 'On farm willow is a powerful bet', article in *Farmers Weekly*, 8 July 1994.
7   See note 5.

CHAPTER 11

Getting Started

No farmer or landowner is likely to say 'I'm going in for energy crops' on the spur of the moment. A new, mainstream arable crop is going to mean large investments of cash, management time and education. Even more difficult, farmers and landowners are faced with a choice: that of producing biomass as a commodity on contract to a large electricity generator, or of creating his or her own market and then making an even bigger investment in heat or power generating equipment. However, here in the United Kingdom you would not be entirely on your own because the British government and the European Commission are offering incentives and help — at least in the understanding and knowledge of biomass crops such as willow, poplar and miscanthus (or elephant grass, another possible biomass crop for the future).

One of the best places to make 'tentative enquiries' is at the annual Royal Show. A site at the National Agricultural Centre at Stoneleigh has been set aside as a demonstration area for short-rotation coppice. It is suitably placed between the forestry demonstration area and the agro-forestry area. There are various types of coppice being grown there and at Royal Show time there are many stands with equipment, sales literature and advisors on hand to discuss all aspects of the business. Also, every two years, the Royal Agricultural Society of England, British Biogen and the Wood Energy Development Group hold a conference to discuss the latest aspects of the industry, at Stoneleigh.

Government, through ETSU, has helped set up five (with an independently run one in Devon, making six) demonstration farms, across the south of England. From time to time these farms hold open days, when you can visit, see the crops, discuss possibilities with the farmers and then come to your own conclusions. Contact ETSU to discover when the next open day will be. ETSU also runs a Renewable Energy Enquiries Bureau, which is available to provide literature and information free of charge. Government too is funding research and demonstrations of energy crops at some of its experimental husbandry farms, such as ADAS High Mowthorpe and ADAS Pwllpeiran. Add to this research stations such as Long Ashton and Loughgall; and private willow and poplar growers like Murray Carter, Lionel Hill, Chris Whinney, Rupert Burr and John Seed and you will find there are quite a few places where you can go to see the crop and to discuss its prospects.

If you feel that the future for you lies in supplying woodchips on a long-term contract to an electricity generator, then contact traders like Banks of Sandy, who will be pleased to discuss possibilities; or talk to consultants such as John Lockhart, Edward Willmott, Major Edward Stenhouse, Dr Clare Lukehurst or others listed in Chapter 15 who have wide experience in this field. For small-scale on-farm production of heat or electricity contact manufacturers and agents for equipment — such as Bob Talbott or the engineers LRZ, Bernard Wilkins or ESD; and of course ETSU has many specialists who are most approachable. Check through Chapter 15 to see who is nearest to you. If you have a group of friends and neighbours who think they might be able to go in for short-rotation coppice for local markets (starting small) then you may like to contact my own company, Home Grown Energy Ltd, listed in Chapter 15.

THE NEXT STEP

Having made a couple of visits, discussed the crop with various people and decided that you are still interested — now is the time to study. This book should have given you many sources which you can pursue. You also need to study your locality to see whether there is a market for what you might produce — or whether there is the possibility of creating such a market. Is your land suitable for short-rotation coppice? Are there fields with good access for tractors and trailers, even in a wet winter?

Where on the farm could you plant this crop — bearing in mind that it will be in the ground for up to 30 years? How would such a crop fit in with existing enterprises? How would it affect your cash flow? This is no small step! A time, perhaps, to discuss the whole matter with your usual consultant — once you have made sure that he or she understands about the crop and, most importantly, has experience of growing it on a field scale.

With a new crop like this it is remarkably difficult to come by facts and figures, especially costs and returns. This is partly because they are changing all the time — with costs falling and potential returns fluctuating almost from day to day. (Energy costs all relate to each other, so coal, gas or oil prices will affect the price of woodchips.) Each situation is different: each plan will be unique because your situation may be ideal; the offer you get for a boiler may be very good; or a main road could prevent you delivering woodchips without an expensive detour. The answer is to carry out your own feasibility study. If your scheme is a large one, or involves several other farmers, you may be able to get financial assistance for it. If that is the case, then hiring a specialist engineering company to do such a survey will be well worth while — even if you could carry out the work yourself — because financiers and potential customers are more likely to respect the results.

The feasibility study will consider all the possible pump-priming and longer-term grants that could become appropriate, be it set-aside, woodland planting grants, development or marketing grants. It will consider various choices of equipment, obtaining quotes from several possible suppliers, and it will present you with a financial picture of the proposed project: how much capital you will need; your costs of production; and the prices you will need to achieve to be viable.

Your input to this feasibility study will also be important because your local knowledge is what may provide the winning element. For example, the scheme may not prove viable unless you produce heat and power for your own farm and home and for the homes of neighbours and local businesses. Even after all this you will still have to 'take a view' on the world energy situation, the world environment and whether political pressure to reduce carbon dioxide and other emissions is sufficient to give long-term security to the production of home-grown energy. This is, of course, very daunting. You can, however, play for time. You can gently work towards introducing the new crop without huge financial commitment.

At Home Grown Energy — Calne, the group of farmers and the rest of us who are at the 'interest' stage are gathering knowledge and have commissioned a feasibility study with FEC Ltd, a well-respected group of engineers. We have invited visiting speakers such as Murray Carter and Chris Whinney (potential suppliers of planting material), Dr Paul Maryan of ETSU and Dr Adrian Bowles (now with Border Biofuels) and have approached the local district council and school governors to ascertain their attitude to the possibility of us heating their sports centre and secondary school with locally grown wood-chips. So far, all signals are still 'green' so we are proceeding cautiously towards the next step.

LOOK AND LEARN

To give us a point of interest we have a half-acre trial plot of willow. Murray Carter supplied ten different clones; one of the local landowners has supplied the land, the rabbit fencing and cultivations; and we are all tending the crop — despite it being planted late, in a rush, on very wet land (so we couldn't cultivate properly or carry out proper weed control before planting). After one week (in May 1994) the willow cuttings sprouted vigorously. So did the weeds. Consulting Neil Roberts, who is Murray Carter's manager, I discovered that spraying at this tender stage was very hazardous for the willow — so, partly for physical fitness' sake and partly for the willow, I have been weeding the plot by hand, using a two-way long-handled knife hoe from the Henry Doubleday Research Centre at Ryton Gardens near Coventry.

Family and other members of the group have also shared this experience — which has added muscle to our frames but, more importantly, brought us very close to the willow (sometimes much too close, when an unfortunate cutting is struck a glancing blow and on occasions, pulled right out — and pressed hurriedly back). A heavy duty motorised garden rotovator brought relief, weeding between the rows, breaking up some of the clods at the same time. This made weeding between the plants with the hand hoe much easier and enabled us to get on

The author learning about weed control on the half-acre trial plot of willow clones near Calne in Wiltshire.

We should have cultivated more finely, planted earlier and taken sensible precautions against weeds, but we were late, the weather was wet — it was all done in a hurry — and the hand-weeding was very good for me.

top of the weeds for the first time. I should, however, make the point that for larger plots, and on a field scale, suitable herbicides are indeed available for weed control during the growing season. Mechanical control of weeds is also practised in Sweden, with some success.

We watched how the crop developed. We saw how young shoots were nipped off as if cut with a knife. Murray Carter

A rich tapestry of mixed willow clones in trials at Long Ashton Research Station. The larger area at the bottom of the picture is a mixed clone experiment, while the smaller block at top left is a trial for rust by the International Energy Agency. (Long Ashton Research Station)

Candy Stevens, a generation engineer with South Western Power at Bristol, one of the companies which is seeking to establish power stations fuelled by short-rotation coppice. Candy is a member of the executive committee and of the export sub-committee of British Biogen.

The many different bark colours exhibited by willow. (Ken Stott)

Scarlet stems and blue-green leaves of the red willow *S. alba* 'Chermesina'. (Ken Stott)

A selection of leaves from different willows. Clockwise from top left: two groups of *Salix purpurea*; × *dasyclados* (*viminalis* × *caprea* × *cinerea*); *aegypteaca*; *pentandra*; *nigricans*; *alba*; *matsudana* 'Tortuosa'. (Ken Stott, Long Ashton Agroforestry Consultants)

Svalöf Weibull cuttings produced by Murray Carter about to be planted on the demonstration plot at Calne. (George Macpherson)

The 'step-planter' at work. In the background, a stand of three-year-old coppiced willow. Note the quality of tilth needed for planting willow. (Murray Carter)

Mechanical weeding in progress with a finger-weeder. (Murray Carter)

Newly planted willow coppice. It sprouts after just a few days. (Murray Carter)

Caterpillar attack. (Murray Carter)

A fertiliser spreader
adapted for high delivery
of granules.
(Murray Carter)

A Claas harvester at work on young coppice. Short rotation coppice makes use of silage machinery in the winter. (Murray Carter)

The new 'bender' harvester. This could offer a wide new range of storage, drying and conversion systems. (Murray Carter)

Rust disease
on a poplar leaf.
(Murray Carter)

A large willow-fuelled power station in Sweden. (Murray Carter)

thought this was most likely to be slugs — since rabbits and deer were not plainly apparent — so I was given slug pellets to broadcast and that seemed to stop it. Interestingly, those cuttings which were nipped off have sent out several shoots to replace the original one or two. We shall have to watch their progress.

After the initial flush of weeds, we experienced a drought — which was excellent from one point of view, because the weeds died so quickly — but it also hit the less sturdy of the willow cuttings. The fatter ones seem to be better equipped to cope with dry conditions. Also, the poor cultivations showed up. The heavy wet soil had not broken down to form the required seedbed and we were left with cloddy conditions. As the soil settled, some cuttings were left sticking out of the ground and needed pushing in and consolidating to make the most of the damp soil lower down. Luckily we noticed this early enough to do it without damaging too many roots. It would not, however, be economic to go around pushing cuttings down and stamping around them on a commercial plot! Much can be forgiven on a test plot — I hope. It certainly impressed on all of us the need for good land preparation, proper weed control and post-planting consolidation — if you can possibly do it. No doubt when we start preparations for commercial plantations, the weather and soil conditions will do all they can to prevent this.

In retrospect, it would have been a good idea to include poplar in our trial plot. This would have needed a little more land but would have given the opportunity to compare productivity and husbandry on a given soil and in similar moisture conditions. Perhaps it's not too late — the landowner may wish to expand the plot. In future it may even be 'the thing to do' to plant mixed species stands of coppice — one row of willow and one of poplar. That option is being studied with a view to increasing yield and as an extra guard against the spread of diseases and pests. On a field scale we would not have had the luxury of all this time and inefficient weed control: as I mentioned in Chapter 6 the whole venture has to be planned properly and planting should not take place until cultivations and weed control systems are properly in place. There is much to be said for a contractor coming in to establish the crop.

Most grants which are available for pump-priming or for establishing woodland must be approved before any work begins — and if you start before you get the go-ahead you stand a good chance of losing the grant. The wet spring, our late start and the general rush to get things done contributed to our losing out on that front, too! MAFF offers the possibility of a grant through the Marketing Development Scheme — which helps cover feasibility studies and management costs for an initial period — acting as a 'pump priming' mechanism.

It was fortunate that there was a grass meadow near the wood and near the farm buildings that was suitable for our trial plot. Other farmers in the group were reluctant to give up grassland because they are dairy farmers, and arable land because of all the complications of IACS forms. One possible alternative might have been at the sports centre or the school which we intend to heat. In this way not only would we all have learned much about the crop and about the various clones being displayed, but both general public and school pupils could also have taken a closer interest in the whole thing, with co-ordinated studies of the wildlife being carried out by the sixth formers in the biology curriculum.

LET PEOPLE KNOW WHAT YOU ARE DOING

In our group we believe that keeping local people informed at this early stage is vitally important. It is an easy subject to discuss because people are very much aware of atmospheric pollution, of the need for sustainable energy and for restoration of wildlife habitat. I have been asked to speak to several local organisations such as the Rotary Club and Civic Society and I believe this is very important, because if rumours start it may mean that the whole project is put in jeopardy. If people get completely erroneous ideas about home-grown energy, such as that there will be smoke billowing from high chimneys, or large numbers of lorries or tractors and trailers carrying woodchips through the town, or willow coppice stinking of sewage sludge in the hot summer, then it will be hard to get co-operation from local government or companies. It is not difficult to get speakers on the subject of short-rotation coppice if you feel the need for local information from credible sources. ETSU provides plenty of useful literature on the subject, too, and there are some excellent videos about.

FACING THE REALITY

When the feasibility study is complete, and if the figures look as though the pay-back period is acceptable, the real work will begin. When starting up a new enterprise involving the production of energy crops, there will be aspects which will probably be new to most people involved. For a start, there will long-term contracts to be negotiated. The farmers will not want to plant a 30-year crop without having a market to match. The buyers of energy must have reliability and continuity. And both sides of the deal will want a fair and profitable price. ETSU can provide an example of a supply contract for wood fuel.

It is advisable to approach the banks early in the proceedings if you are going to look for finance. The bank's idea of 'risk' and how it will be defrayed may be different from your own. The manager may ask you to put in place certain guarantees relating to the fuel supply side — just in case there's a terrible crop failure or a devastating disease reduces yields. There will have to be a fall-back position — like a standby oil boiler — so it is better to get the bank's specification from the outset so you can follow them. Insurance companies will need to be consulted too, regarding third party risks and the protection of staff operating boilers and machinery.

John Seed of Border Biofuels says that any good lawyer can help prepare contracts to suit both sides; John has spent two years or so getting ready for this and has sets of contracts all ready for use. He is willing to provide samples, for a sensible fee, although he reminds us that most government grants in this area do provide for research and the preparation of contracts; groups may wish to carry out the research and planning for contracts on their own, with government help. John Seed feels this is a good idea, because it means that everyone involved goes through all the points that need to be considered and learns of possible pitfalls.

Risk analysis will be a very important part of contract preparation. 'What happens if we get a two-year drought and you can't supply coppice?' 'What happens if we all get snowed in for six weeks and you can't deliver woodchips?' 'What happens if the Department of Education closes the school?' 'What happens to our income if you close the sports centre down for six months for refurbishment?' 'Should we keep the old oil or gas boilers serviceable just in case we have some kind of failure?' 'Whose responsibility should this be?' 'How do we respond to big changes in the general price of energy?' 'How often do we review prices?' All these questions need to be answered and catered for in the agreements.

Some farmers and landowners are applying for NFFO contracts to produce electricity, initially at a subsidised price. After a few years 'convergence' will occur — that is, convergence of the subsidised price and the current open market price. In other words, the subsidy will gradually diminish and then disappear, by which time the biomass-for-fuel market should be well established. To obtain a contract under the NFFO scheme, would-be contractors must provide convincing plans that they would indeed be able to supply the electricity which they are talking about. This is the 'will secure' test which all schemes must pass. Up until now most of the will secure test has been tailored for sustainable energy sources such as wind, hydroelectric, tidal or wave power. 'Fuel' as such, under those types of production, is less of a risk: either the wind blows or it doesn't — but you don't have to do anything to make it blow, or to make the tides run or the waves roll. You do, however, have to do a lot to produce woodchips — as we have seen in this book. There are, therefore, many risks which must be addressed.

British Biogen, the umbrella organisation which has been set up to look after the interests of both producers and users of biomass, is keen that a rigorous approach to risk assessment of the biomass fuel chain is undertaken. This is to help ensure that 'cowboys' do not succeed with casual applications for NFFO contracts. What could be worse for the industry than for contracts to be awarded and then not fulfilled? Every potential risk and loss needs to be considered and measures to mitigate those risks and losses prepared. Full records of all of these need to be kept. If a fuel supplier is taking full responsibility for a constant supply then his or her ability to carry this risk, without loss of supply, must be demonstrated, says British Biogen. Failure would undermine the whole industry.

THE HARD FACTS

The first essential is to be sure where the fuel is coming from. It will be safest to get letters of intent to supply the total quantity of woodchips required — specifying farmers' names and addresses. There needs to be a firm and pessimistic estimate of yield — even in a poor year, with all kinds of problems such as difficult harvest, disease and bad weather. Does the quantity you expect to receive from a particular source represent 50 per cent of what's there? Or 90 per cent? Could you still get sufficient even in a terrible year? You should get evidence that this has been considered and accounted for.

Everyone must agree as to the quality of the fuel to be supplied — and the method by which this is assessed — with provision for arbitration in case of disputes. How much moisture will be allowed in the woodchips? The range of moisture contents must be defined and proposals made for the treatment or disposal of unacceptable material. If the project is to depend on short-rotation coppice, it will take up to four years before the fuel is available — what provision is being made to supply woodchips for those first four years? How much will it cost — perhaps using forestry contractors to extract and chip thinnings and by-products from local woodland?

When it comes to the storage of woodchips, everything must be properly planned and documented, well in advance. You must be clear as to how the woodchips will be stored, how much, and where; and you need a strategy for how they will be protected from fire, vandalism or weather damage. Losses of various types are inevitable: from weather, from deterioration during storage and spillage in transport. All these must be assessed and built into the calculations. You will also need to know who clears up the mess left behind, after storage. Decisions will need to be made, before the event, on who will collect the woodchips at harvest and carry them to the store; who will collect them from the store and deliver them to the processing plant — and when.

What the purchaser of woodchips will need to know, with great certainty, is how secure the supply will be over the next 15 years. What provisions have been made for when things go wrong? What penalties will be applied for failure in any of the links of the supply chain? And if those penalties do need to be applied, will they have sufficient bite — and will they ensure that the project continues smoothly?

And before NFFO contracts are awarded, British Biogen insists that contract terms and prices should be acceptable to the suppliers of woodchips. They do not want to see farmers and landowners selling woodchips as a commodity, having no power over price. The group which prepared the 'will secure' test proposals on behalf of British Biogen included all sides of the industry — producers, processors and consultant engineers.

PRODUCING A COMMODITY

Several electricity generating and distributing companies have applied for NFFO contracts. It is most likely that at least one of them — such as the Hampshire Biomass Project, developed by South Western Electricity in conjunction with Banks Doltons, who will act as fuel suppliers to the project — will have been successful in its bid. There are other possibilities, too, in other parts of Britain. Yorkshire Water has won the promise of a European grant to establish a large project: that will require a minimum of 2500 hectares of willow or poplar to be planted.

Short-rotation coppice will need to be within 40 miles or so of the proposed power station and already companies have been approaching landowners and farmers with a view to signing them up as woodchip producers on long contracts. This type of contract will suit some people better than others and there will be many variations on them. For example, if you are a farmer nearing retirement or who has developed off-farm business and has little time to spend at home on the land, the power companies are offering a complete service including land preparation, planting, fencing, weed and pest control — right through to harvest, storage and every other possible requirement, without the participation of the farmer or the farm staff. This would boil down to contract cropping.

Other farmers may wish to plant even more coppice than would be needed for such a market, so that they could have reserves, or supplies for a spot woodchip market, which is likely to develop. This view is backed up by Richard Whitlock of the grain and woodchip merchants, Banks of Sandy. He believes that, initially, farmers and landowners will be reluctant to commit large areas of their best land to coppice — 'at present it's a "second-best-land" crop' — but once they see the market build up this attitude will change.

Some farmers have been expressing concern that willow blocks land drains, and whereas those on soils with a high iron content may be less concerned at drains becoming blocked with willow roots instead of iron oxide, others will need to consider the 30-year duration of short-rotation coppice crops. Willow and poplar have the effect of lowering the water table, so is it such a serious problem? Much will depend on individual farms and fields — but, most essentially, on markets.

CHAPTER 12

Setting up Farm Energy Marketing Groups

Producing home-grown energy from short-rotation coppice for large projects is not something you can generally do on your own — except on very big farms. With 15 per cent of your land down to set-aside, though, there's enough capacity on even a modestly sized enterprise to generate heat and even electricity for your own use. One company, LRZ, for example, sells a wood combustion unit which can be used to convert existing oil-fired boilers to woodchip combustion. It is suitable for horticultural or other uses. To succeed on a larger scale, however, home-grown energy must make an immediate impact on the market, UK farming, the general public, and on government both at home and in Brussels. It does not come under the classification of an 'alternative enterprise' like ostrich-keeping or growing borage: the type of project a single entrepreneur can build up . It is a new, mainstream agricultural enterprise suitable for both arable and grassland farms.

Alternatives like ostriches may start with one farm introducing the idea, promoting very high prices for breeding stock until the next phase when the birds are actually killed for meat, leather and feathers. At that stage reality enters the picture and prices for breeding stock fall to reflect the value of the finished product. We have seen it happen with red deer, edible snails, angora goats and other minor alternatives. Farm advisors rightly note that although a few people will have done very well out of such projects during the early development stages, the majority of farmers could not even consider entering such a small market place. On the arable side there are alternative crops like evening primrose, lupins and navy beans, although the prospects of them covering very large areas of land seem limited.

Energy crops are different. Even rape-seed oil for vehicle fuel — which is considered to be an inefficient and expensive way of producing fuel, but which may have niche markets if politicians can be persuaded to waive taxes or offer subsidies — is quite likely to occupy many thousands of hectares of UK arable land. Short-rotation coppice, and the associated requirement for forest by-products and thinnings during the establishment period, will cover many hundreds of thousands of hectares of arable, grassland and forest land. It will involve thousands of farmers and landowners, government officials, engineers, merchants, fence, fertiliser, spray and information suppliers. Indeed, it is all set to become *big business* over the next ten years, for the reasons we have already discussed: it is environmentally very friendly; it is sustainable; it makes use of spare rural production capacity; and it will provide useful income for all concerned. There is a strong case for sharing the risks involved — and the rewards. After the energy minister at the time, Tim Eggar, announced the third tranche of the NFFO he said that 'he expected this to stimulate substantial commitment to coppicing'. The embryonic UK short-rotation coppice industry responded by submitting projects totalling more than 12,000 hectares of planting in three years. If fulfilled, this would exceed the total Swedish planting programme so far — and Sweden is acknowledged as the current world leader in this field.

If farmers and landowners are to reap the full benefits from this new industry they must first make sure they add as much value to their product as possible, *before* it leaves their ownership; and they must work together in units of a size which enable them to exploit the benefits of scale.

A farmer on ten hectares can buy a pair of ostriches and build them up into some useful money in a matter of five years. A farmer on a hundred hectares of land might easily be able to plant sufficient short-rotation coppice to be able to supply his or her own house with heat and the larger houses in the vicinity. But on one farm of this size there simply isn't enough available land to produce sufficient coppice to be able to, for example, sell electricity to the national grid. First, the farmer will not want to commit the whole farm to this one enterprise, so output will be too small for more than a few days' production of saleable power. Second, the farmer will not be able, at this stage, to get equipment of a suitable size for such a small quantity of wood-chips — and 'small' tends to be 'inefficient' in this field.

The argument, therefore, goes like this: 'If you don't want to commit a very large proportion of your land to short-rotation coppice there are various alternatives. You could contract with a merchant or an electricity generating company to supply wood-chips; you could use your small supply of woodchips to heat one or two houses, or a farm enterprise; or you could go in with neighbours, friends and colleagues to make your sup-ply part of a much bigger supply to a local processing plant which makes heat, electricity or both. That local plant could belong to the farmers and landowners supplying it — in which case they would have a much higher capital commitment, as well as management responsibilities; or it could belong to an independent company. In the latter case, as a member of a group of suppliers you may not be quite as exposed as if you were contracted on your own to supply the woodchips as a commodity.'

THE SUSTAINABLE ENERGY PRODUCTION GROUP

The most fertile area in which to start looking for potential partners in a sustainable energy production group is an existing discussion group you might belong to. If you are comfortable in its membership then it is likely that other members think along the same lines as yourself. The subject of 'short-rotation coppice' may be ideal for the winter programme of visiting speakers — and you will find plenty of speakers listed in Chapter 15 of this book. Another place to find sympathy is within a buying or selling co-operative, or within a local branch of the NFU or CLA. As we discussed in Chapter 11, on getting started, the group can take steps to learn more about the subject and then move on to setting up a working party.

This small group can invite other potentially interested parties to participate in one or more meetings. For example, John Seed of Border Biofuels, one of the first to stimulate the setting up of such groups, says 'get hold of the water company people early on in the proceedings'. They are always interested in finding new places to spread sewage sludge. Contact with local environmental groups may also be a good idea — especially if you already have a sympathetic relationship with their membership. The benefit of this is that myths can be quashed and matters of concern (which the agricultural community may not have considered important, but which conservationists feel strongly about) can be addressed. If local people can follow the progress of your plans and understand the motives and likely outcome, there is less chance of nasty shocks on both sides, with consequent wrangling, planning enquiries or even boards of enquiry.

In Denmark, for example, where groups of local people became involved in schemes for wind farms, there have been far fewer disputes and court cases: local people not only understand the details of the wind farm but they also tolerate its disadvantages (such as low levels of gearbox hum) because they themselves are benefiting — through land rent, part-time jobs and part-ownership of the capital equipment. Schemes instituted by 'big money' from the outside are more likely to come up against the local people because the only consequence of having wind farms in the vicinity is then seen as loss of scenery and an increase of irritating noise. As the saying goes, 'It's always other people's pigs which smell.' Local ownership and participation are likely to speed the development of home-grown energy.

A LOCAL EXAMPLE: BORDER BIOFUELS

In north Northumberland and in the south-east of Scotland, Border Biofuels is showing how groups can make fast progress unattainable by the individual. What prompted John Seed, the instigator of the project, and his colleagues to start their activity was a report by the Rural Resource Management Department of the Scottish Agricultural College in Edinburgh. It highlighted the changes that would result in the Scottish Borders area from the reform of the Common Agricultural Policy. The report suggested that the cereals sector would lose more than £2 million a year and employment levels in agriculture would fall by 10 per cent. Mainly arable farms would suffer a severe cut in income. Indirect effects on associated businesses would also be dire: the grain trade was set to lose £6.6 million a year; the supply trade £1.7 million, and machinery would probably fall by £1.3 million a year. Set-aside would no doubt increase and many farmers would have to find something else to grow.

Energy crops looked the most promising, so John Seed and friends began to explore potential markets for biomass, contacting many organisations, including ETSU and the Association of Independent Electrical Producers (AIEP). It became plain to them that the key to establishing a market for woodchips lay in supplying electricity to the national grid. To do this they would need to obtain a licence under the terms of the NFFO, so as to create a demand for a large tonnage of woodchip which would enable biomass production from arable land to develop as a viable alternative crop for farmers like themselves.

They formed Border Biofuels in the spring of 1992. The first need was to demonstrate their ability to produce commercial quantities of clean power using existing, conventional forestry by-products. They approached ETSU for help and the unit employed Aberdeen University Wood Fuel Research Group to examine this resource. It transpired that the quantities of wood fuel that were available locally and the pattern of fuel supply over the next ten years could be worked out. Using this information Border Biofuels worked out where the best place to build a power station was, in terms of haulage and general logistics, both for the existing resource and for the developing biomass supply coming from short-rotation coppice on arable land.

There was sufficient fuel available for a plant of about 5 megawatts of electricity. ETSU employed consultant engineers to identify suitable steam turbine-based generators and compare costs and various systems. The study concluded that the plant would be viable if there was a customer for the heat which would arise as a by-product, or if there was enough cooling water freely available for the condenser. This was no good for Border Biofuels, because it wouldn't leave a high enough price for the woodchips from forest residues, or encourage farmers to plant short-rotation coppice. They had to find a better way of processing the woodchips and they turned their thoughts to gasification. A prototype gasifier went on public view at the English Royal Show at Stoneleigh in July 1994.

'Having found the "machine",' John Seed told the Chilford Hall, Cambridge conference on 'Wood, a new business opportunity' in December 1993, 'we were able to develop the project further and offer potential woodchip suppliers an attractive price which would enable them to purchase the chipping machinery and other equipment that would be required to supply fuel to outplant. This also meant that we could raise all of the finance necessary to fund this project — around £6 million.'

At the time of writing, Border Biofuels was awaiting the result of the third tranche of NFFO contracts. If everything goes well, John said: 'Instead of fields of redundant arable land we can introduce biomass crops such as arable energy coppice; instead of discarding or burning the residues left after felling operations, we will be able to harvest them as a fuel feedstock for this plant. Apart from the obvious benefits that generating power from this type of renewable resource holds, it also brings substantial long-term economic benefits to the rural community that it serves. Our plant will pay out over £2 million annually for such things as woodchips, transport, wages and local taxes, which means that all of that money will enter the local economy in a 60-km radius around the plant.'

There was also a prospect that Border Biofuels would be able to supply processing plants to similar groups nationwide and internationally. John Seed told delegates that the project had required a great deal of commitment and concerted action from *all* of the players in this new industry — not just the Border people, but also official bodies like the DTI and MAFF, as well as the NFU and commercial interests.[1]

A NATIONAL EXAMPLE: BRITISH BIOGEN

Pioneers like Border Biofuels, national representative bodies and commercial companies have already taken steps to work with each other, too. Take British Biogen, for example. It grew out of the British Wood Energy Development Group — a small bunch of enthusiasts, in government and the private sector — and co-ordinated the demonstrations at the National Agricultural Centre, making a start with representing views and spreading the concept. British Biogen has been expanded from there, to include potential buyers and users of biomass as fuel, or electricity. Although individual members are likely to come into constructive conflict with each other — for example, when agreeing prices — there is sufficient in common between members to enable them to speak with one voice to government, or to their individual industries — farming, landowning, forestry, plant manufacture and electricity production and distribution. The attitude is 'there's enough in this for everyone' and the aim is for everyone concerned to pull in the same direction rather than squabbling over morsels. Membership includes the National Farmers Union of England and Wales, the Scottish NFU, the Country Landowners Association, South Western Power, East Midlands Electricity, and many other companies, farmers and individuals. Already it is being consulted by various government departments for advice on such matters as 'how can government best help projects get established?' and 'can British Biogen make suggestions as to how a NFFO scheme could be devised for heat-only projects?'

The NFU of England and Wales has provided headquarters facilities for British Biogen; and founder members have contributed funds to employ the chief executive, Peter Billins, who has useful experience of opening up new markets. His first task was to co-ordinate the legal establishment of British Biogen as a limited company and to consult founder members on drawing up a constitution. This done, membership was opened up to interested parties, and preparations for appropriate training courses begun. Management of a new crop like short-rotation coppice requires a great deal of learning; and with a large-scale crop in which there are likely to be low margins — at least initially — keeping costs tightly in check will take plenty of skill and knowledge. This can be attained most economically by groups of farmers and landowners — another good reason for not trying to go it alone.

Membership of British Biogen is envisaged as being very diverse. From the national representative bodies of landowners, foresters and growers it extends to agricultural merchants and co-operatives; equipment suppliers; transport companies; utilities and power generators; and construction companies. There will be a role for research organisations; local authorities; property development agencies and companies; heat suppliers and end users; and, of course, professional bodies and associations.

THE FIVE DEMONSTRATION FARMS

When ETSU encouraged Dr Clare Lukehurst to design the plan for 'The Farm Wood Fuel and Energy Project', which was to start in 1991, her brief included working in groups, where appropriate. Edward Stenhouse persuaded Clutton & Clutton to join the project and also played a major role in getting things moving. The aim of starting up five demonstrations of short-rotation coppice, across southern England, was to show farmers how the crop could be grown commercially within a variety of farming systems, and the circumstances which would probably be needed to make it economically attractive to themselves and other farmers. A sixth farmer (Ray Treemer, from near Exeter in Devon) joined as an associate. Some 80 farmers originally applied to be considered for hosting the demonstration, so plainly the idea was already attractive. The Department of Energy (before it was absorbed into the DTI) showed vision and faith in committing large sums of money to this project in an attempt to break the 'chicken and egg' cycle which led to short-rotation coppice being considered as a serious contender for NFFO projects. It should also be said, perhaps, that farmers in general have been very slow to join in with the work on the six demonstration farms. It seems they need more certainty about markets for woodchips and the seriousness of govern-ment intentions for the crop.

The growers started work early in 1992. Cuttings were purchased centrally in the first planting year and competitive prices were negotiated. In the second year growers bought their own cuttings, as well as using planting material from the cutting back of the first year's growth. As Major Edward Stenhouse, who worked with Dr Lukehurst in establishing the five farms, pointed out to delegates to the Cambridge conference in Decem-ber 1993 ('Wood, a new business opportunity'), the individual buying of cuttings showed a price increase over the pre-vious year, 'demonstrating very clearly the power of collective purchase!'

ETSU recognised the need for co-operation between pro-ducers with this new crop and one of the central roles of the farms, which are described as 'growing centres', in the scheme will be to form the core of grower groups selling fuel wood to local customers. Such customers, suggests ETSU, could be glass-houses, poultry farms and dairies; food processors, laundries and factory units; brick, cement and clay works; conference and leisure centres; hotels and schools; churches and community centres; and building conversions and housing schemes. The groups of farmers have the potential for using wood either to generate combined heat and power, hot water, steam or hot air. Dr Paul Maryan of ETSU told delegates at the 1993 conference on short-rotation coppice at the National Agricultural Centre that: 'Combustion systems are already available and have a long reliable track record of use around the world. All that is lacking is a regular dependable supply of fuel wood. It is this gap that the co-operatives aim to fill.'[2]

The five farms programme is due to continue until December 1997 and regular open days and demonstrations are being held on the farms. Training sessions show neighbouring farmers and potential participants in the co-operatives the techniques of growing short-rotation coppice, and seminars are conducted to enable them to discuss environmental and marketing issues.

CO-OPERATIVES, PARTNERSHIPS OR COMPANIES?

If you are setting up a group of growers and potential growers similar to the one we have started in Calne, and have identified one or more markets, you will be faced, as we are, with deciding how to organise yourselves. There are numerous possibilities and there appears to be no blueprint as to what will suit any particular group. For example, it may be that to achieve the best harmony between customer and supplier you need to form a joint company. If you have found a flower or vegetable grower who needs heat for glasshouses, it may be best to have that person on the board. In some circumstances it might be advantageous to have someone from the local water company on the board — if the company has an interest in finding a home for its digested sewage sludge. The question of who to appoint to the board will need careful consideration. Are you to maintain a 'supplier to customer' relationship — which could mean facing competition from other suppliers, after the period of a contract? Or should you seek a 'joint venture' in which the energy users and producers seek mutual benefit from local production and consumption of energy, with partial independence from regional or national suppliers of electricity, gas or oil?

The main reason for setting up a limited company for the particular purpose of handling energy cropping is to reduce the risk of financial disaster. You 'limit' your liability; and if you are likely to be turning over large sums of money this course may be advisable. Running a limited company carries with it certain statutory duties, like filing accounts to Companies House and notifying the authorities of any changes of directorships. If your group is not equipped for this kind of activity then it may be better to form a partnership or co-operative — and different laws will apply. Setting up a limited company should not cost more than £250–£500. Your accountant or business consultant will be able to advise on which course will be appropriate for your circumstances.

References

1   John Seed, 'From conception to reality. The development of a wood-fired power generation project', conference proceedings from 'Wood, a new business opportunity', edited by G. E. Richards, published by ETSU, 1993.
2   Paul Maryan, 'The Farm Wood Fuel and Energy Project', proceedings from the conference 'Short-rotation coppice — growing for profit', held at NAC on 24 March 1993.

CHAPTER 13

The Green Credentials

One of the strongest arguments for public acceptance of short-rotation coppice is its battery of green credentials: it produces one of the world's purest fuels, with least emission of contaminant gases; burning it produces no more carbon dioxide that it captures; it would seem to offer ideal sites for the useful placement of sewage sludge; it provides a fertile habitat for wildlife; it makes use of land which would otherwise be 'set aside'; and it provides work and income for the rural community. Nor does short-rotation coppice threaten ecologically sensitive areas such as heathlands, chalk downs, high moorland grazing, coastal dunes or estuarine margins, because the crop does not thrive in such conditions.

This is not to say that the idea of introducing it will not provoke disquiet from environmentalists and others who have not yet understood it. In talking to local groups you will find that there are many questions that are asked time and again. For example when you mention that, for the first few years, until short-rotation coppice comes on stream with production, you need woodchips derived from local woodland, someone is almost bound to ask, 'But surely, if you are plundering local woodland there is a detrimental effect on wildlife?' The answer, of course, is that when local woodland is maintained properly, with thinning being carried out regularly and undergrowth controlled, there is a much richer plant, insect, bird and animal life. Most of Britain's woodland is in a state of tumbledown abandonment and only profitable outlets for woodland by-products — such as provided by the demand for woodchips — will lead landowners, farmers and foresters to bring the

170

Short-rotation coppice
provides a fertile habitat
for wildlife. (The Game
Conservancy Trust)

woods back to optimal habitat conditions. Salvage of wood-land by-products for chipping, does, however, have to be done properly. There is a danger that the woodland floor can be damaged if using heavy machinery during wet weather, for example, so, as with all farming and forestry operations, work has to be carried out 'in the spirit of good husbandry'.

Another standard question is, 'You may be capturing huge quantities of carbon dioxide while the crop is being grown — but surely you release it all again when you burn it?' The answer is, 'Yes, you do release carbon dioxide but, unlike with coal or oil, this does not produce any net increase of that gas in the atmosphere.' It has been calculated that, in Britain, a million hectares of coppiced willow or poplar could counter a 3 per cent increase in national carbon dioxide emissions. This could stabilise the nation's emissions at 1990 levels at least until the end of the century.[1] The use of biomass for energy is one of the most potent methods of recapturing carbon dioxide from the atmosphere. Something drastic needs to be done, according to climatologists, not only about the current high level of carbon (in the form of CO_2) emissions, which reached 6,000,000,000 tonnes per year (six billion in US parlance) by 1990, but also about predicted future increases. We need to reduce emissions by 60 per cent if we are to stabilise atmospheric carbon dioxide at current levels, but the International Energy Agency now projects a nearly 50 per cent increase in emissions before 2010. Third World energy use has doubled since 1970 and is expected to double again in the next 15 years — and expand sixfold by 2050.[2] At the Earth Summit in Rio in 1992, 106 heads of state or government signed a treaty designed to stabilise the earth's climate by reducing their national greenhouse gas emissions to 1990 levels by the year 2000.

There are many other questions about the effects of atmospheric pollution on the environment, but there is a wealth of scientific research findings that provide answers which quickly win support from even the most protectionist factions. For example, a research project (jointly funded by ETSU and the Forestry Commission) to evaluate energy and carbon budgets for the production of wood fuel from short-rotation coppice plantations, conducted by Robert Matthews and Robert Robinson at the Forestry Commission's research station at Alice Holt in Surrey, concluded that 'Such plantations yield substantial net energy and carbon benefits.' Let us take a closer look at this aspect of the impact of home-grown energy before we consider its effects on the ground environment, water, plants and animals.

The researchers developed a computer model for predicting the effects of different management practices and harvesting systems on the energy and carbon budgets. They took into consideration all the materials and energy used in the manufacture of materials and machinery involved in the production of woodchips. This included preparation of the seedbed, cuttings, maintenance, fencing, chemicals, harvest, storage and utilisation. They even accounted for differences in the efficiency of machinery operators.

They finished up with an 'energy ratio', which represented the total energy value of the wood produced compared with the total energy consumed in its production. In other words, 'energy out' compared with 'energy in'. For wood fuel production from short-rotation coppice to be worthwhile, the energy ratio must show a very much larger balance in favour of 'energy out'. Then, to calculate what they called a 'carbon ratio', they compared 'carbon benefit' with 'total carbon emitted to grow coppice'. One part of the 'benefit' was the total carbon stored in unharvested coppice. Don't forget that if you are working on a three-year harvesting cycle only one-third of the wood will be burned each year — the rest stays in the ground, growing. However, relatively little carbon is stored in unharvested coppice. The other part of the 'benefit' was the carbon emission that was 'saved' by substituting wood fuel for coal, oil or other fossil fuels. So carbon benefit is the sum of 'carbon stored' and 'carbon saved'.

The researchers discovered that much depended on the management systems employed, which could lead to wide differences in predicted energy and carbon ratios. Even where yields of wood were assumed to be very low, however — e.g. 4 tonnes of oven-dried woodchips per hectare per year — the energy ratio was still estimated to lie within the range of 10 to 25 to 1, while the carbon ratio was between 4 to 25 to 1. However, at a much more likely yield of woodchips, say, 12 tonnes of dry woodchips per hectare per year, the estimated range of the energy ratio was not only wider, but higher, ranging from 25 to almost 70 to 1, while the carbon ratio ranged from 8 to 50 to 1. The researchers warned that the delivery costs of the woodchips and energy used in combustion are not represented in these figures, but they were still confident that wood fuel production from short-rotation coppice yielded substantial energy and carbon benefits. The figures could be improved even further, they said, if alternative methods of fencing and new methods of storage of wood fuel were to be worked out and used.[3]

The study did not account for all the carbon stored in the roots of the coppice. This stays in the ground. Leaves are eaten by insects, who are then eaten by birds that excrete their droppings which rot down and build organic matter in the soil. No work has apparently been done to quantify this carbon capture but, as Caroline Foster of ETSU pointed out to the Cambridge conference on 'Wood, a new business opportunity' in autumn 1993, it was likely that the carbon stored in this way would eventually be lost again when the land was returned to high input agriculture.[4] She also reminded delegates that when wood was burnt to produce heat or power, the residual ash — which may total between 1 and 3 per cent of the dry matter — can be returned to the land as fertiliser.

EMISSION SAVINGS IN THE UK

Table 13.1, taken from ETSU's 1994 report, 'An assessment of renewable energy for the UK', shows the emission savings, relative to the average United Kingdom mix of electricity generation plants in 1990.[5]

Table 13.1

Emission	Emission savings (grams of oxide per kWh electrical)	Annual emission savings per typical 5 mWe combustion scheme generating 37 GWh/yr (tonnes of oxide)
Carbon dioxide	734.0	27,000
Sulphur dioxide	8.1	300
Nitrogen oxides	2.4	90

These figures speak for themselves: savings of carbon dioxide and pollutant gases are enormous. Is this important? The Americans certainly seem to think so: Vice-President Al Gore warned (in April 1994) that 'Global climate change is more dangerous to the US than the British navy was to the American colonists in their war of independence.' He said that, 'Our enemy is more subtle than the British fleet. Climate change is the most serious problem our civilisation faces.'[6]

OTHER IMPACTS ON THE COUNTRYSIDE

Transporting woodchips to power stations will have an impact on country roads. For example, a 6 megawatt electrical plant will need some 100 deliveries per week to keep it fuelled. However, as Caroline Foster told delegates at the Cambridge conference: 'Transport movements in relation to short-rotation coppice as an agricultural crop should be reduced, in comparison to traditional agriculture.' She argued that once the

coppice was established the only repeated operations were fertiliser application, pest control (as a last resort only) and harvesting — and that would not be every year, only bi- or tri-annually;
there would be no land preparation, planting and crop maintenance operations. Since short-rotation coppice is harvested in
the winter the total spread of transport movements on and off
the farm would even out, across the year, leading to a lower
weekly rate of traffic; and because coppice is such a low input
crop, transport as a result of deliveries would be reduced.

One possible hazard to the environment that might be caused
by storing biomass such as woodchips was decomposition and
run-off of effluent from a woodchip stack. This is not at all
difficult to avoid: a well-managed stack should dry to about 25
per cent moisture within two to three months. Spontaneous
combustion could take place if a stack were too big and too
wet: this too could be avoided. As for the visual intrusion
caused by woodfuel, while some people fear that short-rotation coppice could become 'just another monoculture', it is not
likely to prove unsightly and, with regard to overhead cables,
there should be no problem with connecting up to the national
electricity grid. Small, wood-fired power stations would need 11
or 33 kV power lines which are suspended from poles rather
than pylons. Different conversion plants will have different
impacts on the scenery. Steam cycle plants are lower and wider,
while gasification plant tends to be narrow and tall. Gasification
is more efficient and requires less land to sustain it but, in
general, the chimney height of processing plant for biomass will
be lower than, or equal to that required for other fuels because
wood has a very low sulphur content and is low in pollutants
such as nitrogen oxides. Wood combustion processes above
0.4 megawatt net thermal input are covered by the Environmental Protection Act with regard to air pollution. Untreated wood
such as short-rotation coppice is not subject to such stringent
emissions regulations as treated wood.

Willow and poplar have the reputation of needing and using
high quantities of groundwater. However, the water requirement of willow grown as short-rotation coppice has been estimated as equivalent to only 500 mm of precipitation: about the
same needed by sugar beet or winter wheat. Short-rotation
coppice plants are capable of utilising groundwater from up to a
metre's depth — the same as sugar beet and oilseed rape, but

less than the 1.2 metre roots of winter wheat. Thus the water requirements of both short-rotation coppice and conventional farm crops are broadly similar, as are the depths from which groundwater can be removed.

The concern expressed as to the impact which the presumed higher water usage of energy coppice might have on water tables or on adjacent crops and hedges appears to be largely misplaced. Coppice plants quickly produce a well-developed system of roots, making efficient use of available nutrients. The roots act as biological filters and nitrate seepage, for example, is less than from arable crops — and will indeed lessen still further as the root system and fallen leaves and debris build up in the topsoil. Nitrate run-off and leaching into rivers and aquifers will be less from short-rotation coppice than from conventional cropping, and soil erosion from wind and rain should also be reduced.[7]

One area in which short-rotation coppice can offer immediate benefit is in 'cleaning up' groundwater. We have already discussed buffer zones along selected river valleys in the UK, where a band of willow or poplar planted between blocks of arable land and a river can stop leaching of plant nutrients. The WorldWide Fund for Nature reported in 'Too much of a good thing: nutrient enrichment in the UK's inland and coastal waters' (April 1994) that more than 112 nationally important wildlife sites are being damaged by excessive nutrient enrichment. It gives examples of lakes transformed from clear water into nutrient-rich soups prone to algal bloom. It blames run-off and leaching of nitrates and phosphates in particular. On a visit to Poland I met a ministry field advisor distributing willow cuttings to farmers to plant around muck heaps and slurry lagoons, to prevent such leaching. Willow or poplar could also be planted strategically to protect waterways, lakes and ponds.

JOBS AND ENERGY SAVING

Home-grown energy and the use of wood-fired power generation plants provides new employment in the countryside.[8] Caroline Foster suggests that a 6 megawatt station would need more than 20 full-time staff, and there would also be an associated stimulation of employment in service industries locally — for fuel delivery, agricultural and forestry contracting, and the supply of equipment and inputs. There may be an important role for short-rotation coppice to play in conserving energy while it is growing as well, by 'modifying the microclimate'. This means that, as part of larger woodland and environmental schemes, the coppice will have unexpected benefits: for example, in acting as a windbreak. This helps protect homes from heat loss caused by wind, thus conserving energy. Graham Hunt, a director of the Forest of Mercia, in his paper to the 1992 conference at Harrogate, 'Wood — energy and the environment', referred to historical times, up to and including the mediaeval period, when there was much more extensive tree cover. Sites for settlements were carefully chosen to give as much shelter from the elements as possible. Removal of tree cover and other forms of windbreak such as hedgerows and scrub boundaries have had the effect of increasing wind speed at ground level.

Graham Hunt explained that the area in which the Forest of Mercia would be established had been opened up to the weather by extensive coal mining, gravel extraction, industrial activity, urban development and agricultural improvements. The landscape was now featureless and devoid of trees and even hedgerows. Urban expansion, because of mining and industrialisation, had tended to be concentrated along narrow routes which gave settlements lengthy edges, increasingly exposed to the wind. In an area of 24,000 hectares there was less than 3 per cent tree cover. Things have been made worse by well-intentioned attempts to rehabilitate land after mining and mineral extraction. The land has been restored to open farmland, with irregularities of contour smoothed out to make it easier to farm. Tree planting, before the designation of the Community Forest, had received a low priority from the restorers. Modern road design helped the wind to whistle into town and village centres. One of the aims of the Forest of Mercia is to put back some of the shelter once provided by trees — and short-rotation coppice could be ideal for this. 'Extensive planting in the open countryside and transport corridors will increase the surface roughness of the terrain and help to slow down the wind stream,' said Graham Hunt. 'Planting blocks and shelterbelts can also be used to deflect wind away from sensitive areas. Within the built environment trees will provide more local wind control and shelter. It is this holistic approach that will enable weather patterns to be manipulated and energy conservation to be achieved.'

MORE WILDLIFE HABITAT

Although densely growing short-rotation coppice does not encourage the growth of some plant species 'under its feet', it has been shown that, over time, there is a development of non-competitive flora under the crop canopy. This is desirable because bare soil is more subject to water erosion. The under-storey of plants develops where excessive chemical use is avoided: invasive weeds are driven out by the shade from the crop and the ground species commonly found under traditional coppice begin to move in. Also the necessary access rides, between blocks of coppice and on headlands, do provide a surprisingly large area for wild flora. These 'corridors' to allow harvesting machinery, sprayers and manure spreaders access once every three years or so could provide up to half a million hectares of new wildlife habitat by the year 2010, according to the government's interdepartmental report Energy Paper no. 62, published in March 1994.[9]

Of course, much will depend on how these corridors are managed. Some landowners may wish to make commercial use of them — as special areas for horse riders, or as nature trails, and will manage them appropriately. Others may be keen ornithologists and may just keep the rides seeded with wild flowers and then mown or topped annually, to help provide good forest-margin hunting ground for raptors. Along these rides, according to surveys carried out, the presence of field voles, bank voles, shrews and bats is likely to increase.[10] Once every three years or so the rides are likely to be churned up by forage harvesters and trailers, but not over the whole area — just on that part being harvested. Restoration can be carried out in the spring.

Current research work is showing that short-rotation coppice represents a semi-stable habitat which re-creates woodland edge conditions on farmland which may formerly have been completely open. Since the crop is not treated annually with pesticides the wildlife populations in and around the coppice can develop; once the crop canopy is well established it smothers out weeds, while pests, so far, have rarely been a serious problem. Work in the UK is supported by observations in Sweden and the USA which suggest that, as the crop develops, a range of species disperse into it. This development of stable wildlife populations in and near the coppice also allows coppice managers to utilise natural systems to control pest and disease attack — for example, by establishing relevant flora and fauna around the coppice edge or in the under- storey. In its turn this should minimise or negate any need for chemical control, in the same way as do 'beetle banks'.[11]

The crop itself attracts all manner of wildlife. I am told by scientists at Long Ashton that nearly 900 species of insects have been associated with willows, primarily Lepidoptera, Hemiptera, Hymenoptera and Coleoptera. According to some authorities our native willows support more insects and mite species than any other genus of native tree or shrub.[12] One study has shown that the conversion from arable cropping to short-rotation coppice significantly increased some invertebrates, notably earthworms but also harvestmen and woodlice, although it did decrease the numbers of spiders and beetles. Four years after establishment the coppice plots contained 145 plant species while the arable plots had only 17.[13] Willow is one of the first plants to flower in spring, attracting not only bumble bees and providing them with food but also the general public in search of 'something green and attractive' for their flower arrangements.

THE BIRDS

One survey of bird populations — which are a good indicator of species diversity — in mature short-rotation coppice recorded more than 30 species with 10 resident, 12 migrant and 16 breeding. Eighteen species were insectivores. A more extensive survey carried out from 1978–90 showed that a coppiced natural willow stand of 875 by 125 metres supported 27 breeding species at a density equivalent to 1491 pairs per square kilometre.[14]

Yorkshire coppice cuttings producer Murray Carter reported that in 1991, on 0.63 ha of one-, two- and three-year-old coppice bordered by hedgerows, counts of birds were taken by local ornithologists, using bird netting. They counted 308 Hirundinidae; 159 Sylvidae; 117 Turdidae; 47 Fringillidae; 28 Paridae; 21 Prunellidae; 15 Aegithalidae; and five others. The list of birds included blackbird, blackcap, bullfinch, chaffinch, chiffchaff, dunnock, greenfinch, redstart, robin, sand martin, sparrowhawk, swallow, song thrush, blue tit, coal tit, great tit, long-tailed tit, willow tit, sedge warbler, willow warbler, garden warbler, whitethroat and wren.

Rufus Sage of the Game Conservancy Trust carries out some research in one of the rides among short-rotation willow coppice.

Mr J. Wilmer of Friars Court in Oxfordshire recorded many more than this on a 4 hectare site of one-year-planted coppice, bordered by a river on one side, between May and July 1992. He adds to Murray Carter's list coot, crow, cuckoo, curlew, stock dove, turtle dove, tufted dove, goldfinch, Canada goose, black-headed gull, grey heron, jackdaw, kingfisher, lapwing, linnet, little owl, magpie, mallard, house martin, moorhen, grey partridge, red partridge, pheasant, pochard, redshank, reed-bunting, rook, skylark, starling, mute swan, common tern, mistle thrush, treecreeper, yellow wagtail, whinchat, lesser whitethroat, wood pigeon, great spotted woodpecker and yellowhammer. Apparently, short-rotation coppice is not likely, in itself, to offer a good breeding site for most birds but they can breed nearby — in hedges and woodland — and use the coppice as a feeding and roosting site.

The Game Conservancy Trust at Fordingbridge in Hampshire has carried out research which suggests that short-rotation coppice may provide a suitable alternative to traditional cover crops for game such as kale or maize. The crop needs cutting more frequently, to keep it low and bushy. The conservancy produce an 'index of holding capacity' for game birds in different habitats and, comparing mature beech woodland with willow coppice, the index for the beech wood is 7 game birds per hectare, while for coppice it is 42.[15] In Denmark small willow plantations are often planted with no other commercial purpose than game management.[16] If you are keen to study this subject further, I recommend another ETSU report called 'Enhancing the conservation value of short-rotation biomass coppice — Phase 1, the identification of wildlife conservation potential', prepared for ETSU by the Game Conservancy, in 1994. It is bursting with fascinating and useful information.

WILL IT OFFEND THE EYE?

'Large areas of arable energy forest will naturally have a visual impact, although careful siting of the plantation should minimise this,' Caroline Foster of ETSU told the 1992 conference at Harrogate. 'Even so, the perception of this impact may well be highly varied. As the crop is deciduous and planted on a rotation it will be continually changing. It also grows quickly and will thus "green" the landscape rapidly.' She believes that the stability given by a long-life crop such as short-rotation coppice may well be very attractive to the public, especially with the change in colours between the seasons. Linking the crop with established woodland may further improve the wildlife potential — but she feels it is important to remember that the crop does not look like woodland. It is definitely a farm crop, which just happens to consist of trees. Visually it is quite different from woodland and people should not be given the idea that 'we're planting a forest'.[17]

References

1 F. Pearce 'All gas and guesswork', *New Scientist*, 30 July 1994, p. 14.

2 F. Flavin and N. Lenssen, *Power Surge: Guide to the Coming Energy Revolution*, Worldwatch Environmental Alert Series, published by W. W. Norton & Company, 1994.

3 R. Matthews and R. Robinson, 'Biomass for energy: progress in computer modelling of energy and carbon budgets for wood fuel production from short-rotation coppice', proceedings from conference on 'Short-rotation coppice — growing for profit', NAC, 24 March 1993.

4 Caroline Foster at 'Wood, a new business opportunity' conference, proceedings published by ETSU after the Cambridge conference, 13–14 October 1993.

5 ETSU, 'An assessment of renewable energy for the UK', HMSO, 1994.

6 G. Graham, 'Gore warns over greenhouse gases', *Financial Times*, 22 July 1994.

7 G. Hunt, 'Energy forestry in the forest of Mercia', ETSU report B/W5/00241/REP, 1993.

8 See note 4.

9 DTI, 'New and renewable energy: future prospects in the UK', Energy Paper no. 62, HMSO, March 1994.

10 B. A. Mayle and J. Gurnell, 'Edge management and small mammals', in *Edge Management in Woods*, Forestry Commission, 1991.

11 See note 4.

12 C. E. J. Kennedy and T. R. E. Southwood, 'The number of species of insects associated with British trees: a reanalysis', *Journal of Animal Ecology*, no. 53, 1984.

13 F. Makeshin *et al.*, 'Short-rotation plantations of poplars and willows on formerly arable land: sites, nutritional status, biomass production and ecological effects', fifth EC conference on biomass for energy and industry, Elsevier Applied Science, 1989.

14 J. Wilson, 'The breeding bird community of managed and unmanaged willow scrub at Leighton Moss, Lancs,' Royal Society of Edinburgh willow symposium, 1990.

15 See note 4.

16 K. H. Nielsen, Handbook on how to grow short-rotation coppice, Uppsala University, 1992.

17 See note 4.

CHAPTER 14

The Way Ahead

There are several questions to consider. Will energy cropping ever take off? And if it does, how will it develop? As with every agricultural development, as much will depend on the weather as on the politicians. Everyone knows that the world's population is expanding fast; but not everyone would agree that existing agricultural capacity can feed that population. I refer you to the beginning of this book, where research units are predicting that production from as much as 100 million hectares of land in the EU will be surplus to food requirements. Will it? What about the Chinese? They would appear to be having difficulties already in feeding their billion mouths. The Worldwatch Institute in Washington is predicting mass hunger within just a few years. Will this mean a bonanza for efficient food producers? How does one take a view of future trends?

The mood and business confidence of arable farmers swings with their income and prospects. In the summer of 1994 they were confidently predicting the end of set-aside as world demand for grain looked likely to increase — and, indeed, the European Commission was considering a temporary reduction of set-aside. The buoyant mood was, perhaps, because of a good harvest (for many) and rising grain prices — linked with the receipt of very large compensation cheques from Brussels. Tractor sales boomed. A hot spell had pulled down harvest yield predictions for eastern Europe and news from former communist countries was still about the problems of privatisation and investment. Yet even Serbia produced a large grain surplus, despite UN fuel and fertiliser embargoes, and Romania hailed its 20 million tonne crop as a sign of better things to come.[1]

The arguments put forward by economic forecasters would seem to hold: former communist countries have some way to go before their agricultural industry is re-established, but they have huge areas of excellent arable land and if modern developments from the west are adopted, their productivity will soar. It is likely that — except in war-troubled areas — they will be much more than self-sufficient in food within, perhaps, five years. The Chinese programme of industrialisation is already producing cash and the logical place for them to buy grain is from as near as possible, viz eastern Europe, where production costs will still be comparatively low. Given a couple of good seasons, worldwide, grain prices should fall. As GATT slowly begins to bite and subsidies dry up, farmers in the EU will find themselves in a truly global market place. Will prices go up, or down? If they go down, interest in energy and other non-food crops will grow. If they go up, food production will remain favourite.

There is nothing unusual in this kind of situation: whether to plant more cereals or not, whether to increase breeding sow numbers or whether to buy more land. Farmers face it all the time and the classic response in, for example, grain marketing is 'sell some and keep some'. Sayings like 'up corn and down horn' illustrate the kind of risks in farming and help bring about farmers' response to them. Mixed farming, in the past, has helped spread the risk and has allowed new enterprises to come and old ones to fade away, without putting the farming family out of business. Single enterprise agriculture became very profitable during the artificial markets of the 1970s, fanned by subsidy with Britain entering the European Community. We saw 'rotations' such as 'continuous cereals'. Such 'mono-enterprise' thrives in the dairy sector but one is tempted to wonder whether it will continue once the protection of quotas is lifted (as it surely must if EU dairying is not to become fossilised).

So, whither energy cropping? The need for the world to make use of fuel which does not alter the climate will not go away. Commitments have been made by governments to reduce emissions of carbon dioxide and other greenhouse gases. Public awareness of the fragility of our environment is growing — not only in the EU but worldwide, as living standards rise. The arguments for UK farmers to adopt energy cropping as a new farm enterprise are as strong as ever. It would seem sensible for farmers to consider the evidence, look at their own land and see whether short-rotation cropping would fit in well with existing operations. If it would — and before even considering any planting — farmers will need to look over the hedge for potential markets: producing electricity for the grid, heat for buildings or complexes of development or combined heat and power for farm or industrial enterprises. So far, short-rotation coppice seems to have many advantages over other forms of renewable energy sources. This has been recognised by Silsoe Research Institute, ADAS, Rothamsted Experimental Station and the BioComposites Centre. In its report 'Towards a UK research strategy for alternative crops', published in July 1994 by Silsoe, coppice energy crops are given top priority for potential new markets.

MORE ACCEPTABLE THAN WIND OR STRAW?

One thing which became apparent during 1994 was that wind farms are unlikely to be allowed in many parts of the UK. District councils and vigorous pressure groups (sometimes supported by vested interests in other power sources) have ignored the advice of environmentalists, county councils, rural development specialists and ETSU and have rejected planning applications for wind farms on the grounds of 'visual intrusion and noise pollution'. They have been backed by planning inspectors at appeal. In the UK, unlike the Danish pattern, most wind farms seem to be financed and managed by 'big outside companies', and the only involvement of local people has been in landownership. In Denmark local villagers, farmers and industrialists have all apparently had the opportunity to take part in the planning, financing and running of the wind farms, with a consequent reduction of sensitivity to the small disadvantages which this technology seems to present.

The North Devon District Council turned down the scheme by West Coast Wind Farms for 23 wind turbines near Barnstaple, even though the site had no special landscape designation. With the scheme having the backing of the county council and ETSU, and with the site having been described as 'one of the best in the county', Mr Michael Baker of WCWF said:

'This decision means no further wind energy projects will be approved in Devon, one of Britain's windiest counties.'[2]

Energy cropping and local processing plants for power and heat generation should not be as difficult to sell to the public. There are less visually intrusive (if carefully planned — see Chapter 6) and encourage much more involvement from local people both on the production side (in agriculture and forestry), with consumers (especially on heating projects), and with planners and environmentalists. On top of that there are the benefits to wildlife conservation and the creation of new habitats. Production and processing units can be comparatively small (supplying a few kilowatts of power or heat) and can be spread around to be near the sources of fuel. And, as I mentioned earlier in this book, there is much more control over output from wood-fuelled energy plants: you can't 'turn up the wind' — you just have to collect it when it blows.

The public seems to have similar reservations about large straw-to-energy processing plants as they have to wind farms. Where I live, in Calne, there has been uproar about the 20 megawatt processing plant being 'as big as Salisbury Cathedral — with a chimney almost as high!' and even local doctors have protested about the potential ill effects of noxious gases from the plant on their patients. There are complaints that straw will have to be brought from too far to satisfy the needs of this plant. People have worked out how many lorries carrying straw would be needed every day — and then considered their routes. The fact that such a power station would make a useful contribution to improving the atmosphere upon which we all depend, and that steady local jobs would be created, seem to be less important, in the public mind. No local benefits were perceived, except a few extra jobs. Perhaps it would have been more acceptable had local townspeople been offered cheaper electricity?

In a report for ETSU by St Ronans Research on 'Public perceptions of short-rotation coppice' it transpired that people would have few worries about 15–20 per cent of a farm being planted to the crop. The report recommends that coppicing should be 'well regulated' to minimise adverse visual impacts on the countryside; that the public be very well informed of its benefits to the countryside's appearance and wildlife; and that it is a locally supported, integral part of the national energy policy aimed at combating global degradation.[3] Coppice producers will ignore this advice at their peril! It would not be difficult for the public to get the wrong end of the stick. 'Transparency' is the necessary buzzword for development of a new energy industry built around short-rotation coppice. People must know what's going on and how it might affect them. Perhaps they should take part in the planning of new schemes right from the start? Even the choice of crop? At least they should be allowed to understand why any particular type of crop has been chosen, and its likely impact on the countryside.

CROP DEVELOPMENTS

The yield of energy from a unit weight of dry matter is likely to be the same for different crops.[4] One kilogram of dry matter is equivalent to approximately 0.4 kg of oil in energy content. How it is processed and how dry the biomass is will determine how much of that energy is harnessed — but it does mean that there could be a wide choice of crops suitable for the production of energy. Farmers will already be asking themselves, 'Should we go for willow, poplar — or miscanthus — or something else?' The answer is as follows. If you are convinced that short-rotation coppice is 'a goer', then start with willow or poplar, or both. So much work has been done on them over the past 20 years that we know the answers as to how to grow, protect, harvest, process and market woodchips. If you become an enthusiast, so that your interest is not simply economic but obsessive, plant a trial plot of miscanthus and start looking for seed of reed canary grass — and perhaps try out one or two new possibilities for yourself, like sweet sorghum or bamboo! You should be warned, however, that even enthusiasts who are already growing miscanthus are saying that 'we don't know enough about it yet'. Also bear in mind that something like 30 per cent of Britain's trade deficit comprises imports of plant fibre for paper and fabric — the rewards may be great, in the longer term.

Even with short-rotation coppice there are many questions that urgently need answering. For example, could one bring in a cash return sooner by *not* cutting back at the end of the first year, but at the end of the second year, taking a small harvest of woodchips at the same time? Cutting back after just one year must surely upset the plant and you would expect it to have established a firmer root system by year two — although I am told that without the dense mass of shoots produced by the first-year cut-back there is insufficient shading from the crop to keep weeds at bay. Should clones be mixed up as they're planted or kept in blocks or rows? Should willow be inter-planted with poplar, if so, in rows, blocks or what? What is the ideal plant population for this country — or this area of the country? Should it be higher, so that stems are not as thick, making it easier to harvest? Or should it be lower, making heavier trunks with less bark? We need answers to all these questions but this should not prevent anyone from making a start: we already have sufficient knowledge to go ahead and begin business. Let us, however, just look at some future pos-sibilities for new energy crops such as miscanthus and reed canary grass.

ELEPHANT GRASS — MISCANTHUS

Under experimental conditions miscanthus (otherwise known as elephant grass) has yielded more than 24 tonnes of dry matter per hectare in a year in the UK. This was the highest second-year yield in the UK so far recorded and was at ADAS Arthur Rickwood station in Cambridgeshire, on peat soil. In Denmark it has yielded as much as 44 tonnes per hectare per year of dry matter.[5] In Germany it has been estimated that once miscanthus gets into its stride, in about year four, it can produce a steady 25 tonnes a year. The crop takes four years to reach its full potential and a future target of 40 tonnes per hec-tare annually is not regarded as unrealistic.[6] John Kilpatrick, of Arthur Rickwood, is quoted as saying that the theoretical potential yield is 55 tonnes per hectare on good soils in tem-perate regions, but he concedes that little is known about the agronomy, economics of production or the true yield potential of miscanthus if genetically improved types were used. 'We do know the crop will not produce seed,' he says (referring to field crops in the UK climate), 'and will have to be established using plants propagated from rhizome cuttings which currently cost 50 pence each.' At 10,000 cuttings per hectare this is not cheap.

The crop should survive 20 harvests.

Miscanthus is a perennial grass related to sugar-cane and it too possesses the C4 metabolic pathway which gives it rapid growth when conditions are right. It produces annual shoots which reach 4 metres in height, resembling bamboo, although the stems are not as hard. Like willow and poplar woodchips, miscanthus has other potential markets and there are already agents who are selling cuttings and advice. Miscanthus has potential use for paper pulp, chipboard production and as a horticultural mulch, according to Bill Bawden of Rosewarne, the base for Cornwall's LEADER project (that is, 'Liaison Entre Action Development Economique Rurale', an EU project to pioneer innovative ways of sustaining rural enterprises). Its canes may also come in useful to horticultural growers, and its growth is so dense as to make it a useful windbreak or screen.

One supplier, J. J. Harvey, of Kenn near Exeter in Devon, suggests that miscanthus could be of use in reedbeds and soakaways, or for screening slurry lagoons. He says that miscanthus is easy to grow on a wide range of soil types, using existing farm machinery for planting and harvesting. Weed control is cheap and easy; and it can be grown on non-rotational set-aside. At harvest the crop can be brought in as chips or baled. After two or three years the rhizomes can be divided up to obtain more planting material.

At Rosewarne a propagation unit has been set up to supply farmers and researchers interested in growing the crop under their own home conditions. New and vigorous lines of the species *Miscanthus sinensis* and *Miscanthus sacchariflorus* have been developed and plants are available as rhizome pieces or young rooted plants. Although miscanthus will not seed out-doors — which is important because it stops the crop becoming a weed — Marshall Hutchens, who works on the crop at Rosewarne, says they are trying to get it to seed in glasshouses, to enable them to breed better varieties. They obtained 16 different lines of seed from Japan but so far only two are showing signs of being worth bulking up for biomass, although some are showing promise for horticultural ornamental use. 'The plants are very big — you could even say "architectural" and flower arrangers find them very useful,' says Marshall Hutchens. Rosewarne intends to try micropropagation techniques too, in bulking up planting material.

Miscanthus dies back naturally in winter. Leaves drop off, and what is left is the equivalent of a field of bamboo canes. These can be harvested from mid-December until mid-February, after which there is some danger of damage being caused to new shoots. Before Christmas the canes have comparatively high dry matter, but by mid-February they have hardened up even more, making them more difficult to chip. Miscanthus chips could be mixed with willow or poplar chips to give a good burn. To establish the crop soils and seedbed conditions are needed as for willow; diseases and pests are no problem so far, although Rothamsted is suggesting there may be a problem with barley yellow dwarf virus which causes stunting of the leaves and stem.[7] Environmentally, miscanthus seems to provide a good nesting habitat for birds such as wrens, tits, blackbirds and thrushes, says Marshall Hutchens.

Bill Bawden says that *M. sinensis* may be more suitable for lighter, drier ground, while *M. sacchariflorus* may be better in heavier, wetter ground — although he says this is probably an over-simplification and it's generally better to try both species. Rosewarne offers a *Miscanthus Grower Guide* to go with plants (see Chapter 15 for the address).

Research and development is being carried out in several other European countries and an EU Miscanthus Productivity Network has recently been set up to study the crop in a range of soils and climates, including four sites in the UK. In Germany the Veba Oel company is studying the potential of miscanthus as a feedstock for gasification and some pilot combustion facilities using miscanthus as fuel should soon be on line.[8] ADAS has funded a taxonomy project, with the Royal Botanic Gardens at Kew, to sort out the different types of miscanthus

The advice regarding miscanthus was spelled out in *REview*, the magazine of renewable energy.[9] 'Before UK farmers plant miscanthus for energy, a number of important questions will need answering. What species should be favoured? What yield can be expected in this country? How long would a plantation last? What row spacing should be used? Would establishment costs be prohibitive? As with coppice, when and how could an energy market for miscanthus be stimulated? And above all, does this energy crop have the potential to be commercially viable?'

REED CANARY GRASS

When I first heard about reed canary grass I wondered whether willow or poplar — or even miscanthus — could ever compete. This fast-growing grass, native to Scandinavia and other northern temperate zones, lasts indefinitely in the ground, its yields compete with anything, it seems to have no enemies and it has already been used, as a powder, as fuel in Sweden. Better still, you can convert a coal or oil boiler to use reed canary grass just by changing the burner! Listening to Rolf Olsson's paper at the 'Non-wood fibres for industry' conference run by the International Paper Research Association at Silsoe in March 1994, I wondered why I hadn't heard of this crop before. I was told afterwards by other delegates that so far we have not been able to grow it satisfactorily in the UK, although that is not to say that one day the crop may not grace our fields. A start has been made at ADAS Arthur Rickwood in Cambridgeshire, where Colin Speller began some trial plots (with difficulty, I understand — the seed obtained from North America did not germinate well and Colin said a lot of work needs doing to understand why). We should, though, be aware of reed canary grass's existence, as a potential future crop.

Reed canary grass, *Phalaris arundinacea*, is a naturally occurring C3 plant, long recognised as having very high yields. It has been used in Sweden on poorly drained land for forage. As an energy crop the Swedes practise what they call 'a delayed harvesting method'. Unlike willow, which is harvested in the depth of winter, reed canary grass is harvested in spring, once the snow has cleared. This can be as late as mid-May at Umeå — and the moisture has almost completely gone from the crop by then, at only 10–15 per cent.[10] It can then be harvested and passed through a hammer mill to make powder. Alternatively the grass can be baled using high density balers. With this method less plant nutrient is removed from the field than if a summer harvest (for fodder) were carried out. The biomass obtained by spring harvesting is more suitable for combustion because 'the elements unfavourable to the combustion process have been reduced', according to Rolf Olsson. Sulphur content, for example, is nearly halved, from 0.17 per cent of dry matter to only 0.09 per cent and chlorine comes down from 0.56 per cent to 0.09 per cent.

Reed canary grass is commercially viable as an energy crop in Sweden because of high environmental taxes on fossil fuels used for heat production. These taxes, says Rolf Olsson, amount to 160 Swedish krone per megawatt hour (as at March 1994: and that is about £136). He says that the grass has a higher ash content than wood powder and a higher moisture content which makes it less suitable for small powder furnaces: it's more suited to plants of 15 megawatt or more.

OPPORTUNITIES

Every farmer and landowner with an eye to the possibilities of energy crop production must examine his or her own land and the factors affecting it — soil type, climate, farming system and legal designation. Those faced by inclusion of their land in Nitrate Vulnerable Zones, for example, may find short-rotation coppice of interest. Government announced plans for 72 of these sites, where there will be compulsory restriction on the application of nitrogen — without specific compensation. The restrictions will come into force between the end of 1995 and December 1999. These sites will cover 1.6 million hectares in England and Wales, and other such plans are in hand for Scotland and Northern Ireland. They are to comply with the EC's nitrate directive which aims to protect water from nitrate pollution. The compulsory measures proposed include a limit of 210 kg per hectare of nitrogen for organic manure; a closed period for slurry spreading on grass, between 1 September and 1 November, and on other land from 1 August to 1 November. Records will have to be kept of all fertiliser application.[11] Short-rotation coppice may well offer some solutions in such situations.

WIDER MARKETS

In the UK we are 'up among the leaders' in the technology of biomass production for energy. On the international market there are many opportunities for the expansion and development of this technology, and government has indicated that it wishes to encourage such activity. Ahead of us is Sweden, where arable coppice is already in use to fuel power stations and district heating schemes. In North America woodchips are used directly by the wood processing industry.[12] At home there may not be many opportunities for installing district heating systems because our winters are simply not long enough — but there will no doubt be situations where a juxtaposition of a new housing development and a sports centre, fish farm or glass-houses could make it a possibility. Once we have some schemes running in this country the potential for exporting know-how and equipment is great.

The Ukraine provides an example. During the winter of 1993–4 the energy minister warned that the nation had reserves of energy for just one month — and being dependent on imports of gas and oil he called on everyone to cut back on gas by 30 per cent, while factories had to cut energy consumption by half. Ukraine depends on Russia for about 90 per cent of its oil and gas, and is deeply in debt for it.[13] The country has large reserves of coal — some of which it exports to the UK — and Ukraine's coal pits contribute 3 per cent of the world's emission of methane, a greenhouse gas associated with global warming, which also makes it very dangerous to mine the coal.[14] Home-grown energy would appear to be the long-term solution to their problem. Latvia imports much of its energy from Russia as well, and yet Latvia appears to have more than enough arable land, with excellent soils and very high availability of water. It has vast untapped forests and would seem to be an ideal candidate for growing its own energy, providing work and prosperity for its farmers and foresters and improving its own balance of payments.

Biomass power offers what the energy campaigner Walt Patterson describes as 'an imaginative and constructive approach to tackling one of the most pressing international problems: that of surplus agricultural production in industrial countries' by creating the opportunity for farmers to grow fuel. 'Biomass power may therefore attract support from two quite different lobbies in industrial countries, both already in existence and both influential: the power plant and turbine manufacturers; and the farmers,' he writes in *Power from Plants*, his report from the Royal Institute of International Affairs.[15] Since biomass power stations need to be built and run near their source of fuel, they can play 'a crucial role in rural and regional development in both industrial and developing countries'. In an almost poetic way he compares traditional power stations with the new ones, fuelled sustainably:

> A traditional fossil-fuelled power station, which can bring both technology and fuel from a long distance away, may look and operate much the same wherever it is sited; in a sense such technology is detached from its surroundings, exemplifying the concept of technology 'conquering nature', forcing it into a homogeneous mould. By contrast, biomass power is more 'site-specific': because the locality immediately around the power station supplies its fuel, the scale and type of technology used in the station must be suited to the local fuel supply, and thus to the characteristics of the locality — the land, the water, the climate and so on. In this respect biomass power, like other renewable energy technologies, exemplifies the concept of working with nature in all its diversity.

Walt Patterson also makes the point that demonstrating biomass power technology in industrial countries may also encourage its adoption in developing countries. This could lead to industrialisation of Third World countries without the accompanying degradation of the atmosphere. Indeed, the 'Global Environment Facility' administered by the World Bank has already begun to support biomass power development. This type of energy production will not appear attractive to developing countries, however, if it has not already shown its worth in the most advanced countries.

Many developing countries seeking to expand their electricity systems may be able to use their own resources for the purpose, provided these resources are managed sustainably and provided that the countries have access to the appropriate technology. Walt Patterson believes that the World Bank could play a key role in this development. Furthermore, the availability of 'modest amounts' of electricity locally in rural areas of developing countries, for irrigation, could boost local agricultural output of both food and energy crops. Home-grown energy could be instrumental in creating its own energy supply while at the same time benefiting food production. UK companies such as those already involved in short-rotation coppice production and utilisation will also have much to offer developing countries and indeed some of those companies, like ESD Ltd (see Chapter 15) have already been involved in the design and building of biomass-fuelled power stations in the tropics. If you are looking for inspiration, read Walt Patterson's book!

Technology may also open up new markets for cereals, wood-chips and other biomass fuels, and it might soon be possible to obtain vehicle fuel from biomass using the technique of pyrolysis. This is described in ETSU's 'Assessment of renewable energy' report as 'the thermal degradation of a feedstock in the absence of an oxidising agent to produce gas, liquid or char. By heating slowly to a low temperature the proportion of char is increased. If the rate of heating is increased, the fraction of liquid and gas produced is raised. At the extreme, this is called flash pyrolysis. The pyrolysis oil may be used to fuel internal combustion engines or it may have value as a chemical feedstock. By this route it is possible to convert whole crops, including the straw, in the production of liquid biofuels.'

In the USA nearly one-tenth of the nation's motor fuel is a mixture of 90 per cent petrol and 10 per cent ethanol. The ethanol comes from maize grain and its inclusion has been encouraged by tax exemptions.[16] In Brazil pure ethanol distilled from sugar-cane provides fuel for about a third of the nation's cars. Even though the costs of such fuel have fallen drastically since the late 1970s, ethanol in the mid-1990s costs more than double the price of petrol distilled from mineral fossil oil. The Worldwatch Institute in Washington believes that the technology for producing this fuel will improve, although there is little chance that ethanol from maize or sugar-cane will be economically viable in the near future. And, say the authors of the institute's book, *Power Surge: Guide to the Coming Energy Revolution* (which I recommend highly as being readable and most informative), the main limitation to conventional biofuels (such as ethanol) is the shortage of cropland.

'Globally', they say, 'there is not enough grain to fuel the world. If the entire world maize crop were converted to ethanol, for example, it would meet only 13 per cent of current global gasoline demand; the world sugar crop could meet only another 7 per cent. Moreover, most of today's crops depend heavily on fossil fuels for fertiliser, pesticides and the energy that goes into ploughing, irrigation, harvesting and processing. Using those crops as an energy source would yield little if any net reduction in fossil fuel dependence or carbon emissions.'

Worldwatch suggests that short-rotation coppice and other lower grade and less valuable plant materials such as cellulose and other complex organic matter could be converted into carbohydrates and then into alcohol. The US National Renewable Energy Laboratory has developed a process, using enzymes, which can turn wood into ethanol. The cost of this ethanol was reduced from $7.10 a gallon of petrol equivalent in 1980 to as low as $1.30 by 1994. Research workers believe that even that price can be halved, bringing ethanol fuel closer to the wholesale price of petrol. According to one set of theoretical estimates cited by Worldwatch, such technology could allow biomass to provide 38 per cent of the world's liquid and gaseous fuels and 18 per cent of its electricity by 2050 — a figure which is equal to more than half of the world's current primary energy use. But the Institute warns that biomass energy is only as sustainable as the methods used to produce it.

In the USA a group of electricity utility managers, government officials, producers and environmentalists formed the National Biofuels Round Table to come up with principles to guide the development of environmentally acceptable biomass energy systems. From all accounts the ten meetings they held generated a lot of heat! Ralph Overend, Principal Scientist of the Renewable Energy Laboratory in the USA, told delegates to the 'Wood Fuel — the Green Debate' workshop at the Old Palace in Hatfield in October 1994 that the participants in the round table talks had agreed to appoint an independent chairman. That proved difficult and in the end they finished up with a 'conflict resolution moderator'. After about three meetings she brought the meeting to face the question of how producers of biomass could make a profit while at the same time meeting the limitations of environmentalists and the market requirements of the energy system. In the end, there was agreement on all but a few issues; and they would all work together to promote the general principles of biomass production for energy and its use.[17]

CONCLUSION

Given potential markets, suitable soils and countryside into which energy crops will blend — and given that now is the time to invest set-aside money in something that will go on producing for many, many years after set-aside is withdrawn — then short-rotation coppice would seem to have a rosy future. There looks to be sufficient evidence to at least encourage farmers to make a start on bringing in this new enterprise, and the logical way to do this is to join with other local farmers in identifying markets and supplying them, setting up a trial plot of willow and poplar and learning how to farm the crop — initially on a small scale. If or when the time comes to expand, they will be poised to do so.

Reading this book you may have got the impression that 'everything is changing all the time', but Dr Paul Maryan of ETSU has pointed out that energy prices are actually reasonably stable. 'Our gas and electricity bills remain remarkably constant, as do energy sources supplied to long-term contract,' he says. 'Indeed, one of the attractions of energy crop production is the fact that people are willing to enter long-term contracts with possibly frequent review periods to take in to consideration fluctuations in world prices.'[18]

Short-rotation coppicing may be one answer to the need for a profitable mainstream crop; to reducing emissions of greenhouse gases; to replacing some wildlife habitat; to offering a useful repository for sewage sludge; and to creating jobs in the countryside. But developing it is no easy task. To quote the senior civil servant who oversees the renewable energy projects at the Department of Trade and Industry, Richard Kettle, it's not any easy option. He told delegates to a seminar at ETSU that 'The "Energy crops and agricultural forestry wastes" end of NFFO 3 is not for the faint-hearted. Successful developers are likely to have devised novel approaches to the problems they face and secured the long-term commitment of those who can make the project happen.' He also said that projects which initially needed forestry by-products as fuel (until the short-rotation coppice came on stream), were unlikely to succeed when they had to compete for these by-products, if there were other markets open to the forest or woodland owner. 'You will need to emphasise the non-price attractions for the energy market for these wastes,' he said. These might be the benefits to habitat and to the environment in general, especially if part of the deal was to leave the woodland neat, tidy and ready for more efficient timber production. He advocated vertical integration for electricity producers so that they could 'retain effective control of fuel supply'. The same advice may well apply to farmers and landowners: using their own woodland resources to get them over the first fuel-hungry years of the project. They might also be wise to take the advice at the marketing end of the project, too — by selling electricity and/or heat rather than just fuel as a commodity.

Home-grown energy, like most concepts in farming, is by no means new. Until the 1940s very large areas of farmland were devoted to the production of energy, which was processed on the farm, with surpluses being exported to markets. The arable fuel was oats, which was complemented by grass, clover and herbs (including a fair few weeds) and the processors were heavy horses. In the years between 1920 and 1929 there was 797,595 hectares 'set aside' to produce 1.4 million tonnes of oats to feed nearly 777,666 heavy horses.[19]

If you believe, as I do, that a return to sustainable energy production from agriculture is due, don't just wait for it to happen. In the words of Yorkshire short-rotation coppice grower and biomass resurgence pioneer Murray Carter, 'A market of this kind doesn't just happen, it has to be built.' The foundations have been laid by government, funding scientists to discover and assemble the necessary facts about potential energy crops and officials to promote the idea and 'prepare the site' by way of public surveys and case studies; by academics like Professor David Hall of King's College London, who has campaigned for decades worldwide promoting the concept of fuel from the land; by farmers and engineers seeking new mainstream enterprises; by the farming and national media keeping us informed; by chartered surveyors seeking ways forward for their clients; by environmentalists and conservationists trying to bring rationality to modern society; and by plain enthusiasts who can see the good sense in handing over the world, intact, to our grandchildren. Now it's time for the investment in learning, land, money and hard work to set in motion a virtuous cycle of renewable energy.

> Sustainable development is that which meets the needs of the present without compromising the ability of future generations to meet *their* own needs.
>
> John Seed.[20]

References

1 C. Horseman, 'Climate change worries for central European agriculture', *East European Agriculture and Food Monthly*, no. 144, September 1994.

2   'Devon wind farm plans rejected', *Financial Times*, 16 August 1994.

3   R. Sadler, 'Public perceptions of short-rotation coppice', St Ronans Research, ETSU, B/W5/00340/REP, 1994.

4   A. Grimm and A. Strehler, 'Harvest and compaction of annual energy crops for heat generation', in *Producing Agricultural Biomass for Energy*, a report of the European Co-operative Network on Rural Energy workshop, published by FAO, Rome, 1987.

5   C. Speller, 'The production of miscanthus under UK conditions', PIRA conference at Silsoe, 23 March 1994.

6   'Elephant grass has a big potential', *Farmers Weekly*, 22 July 1994.

7   R. Plumb, Crop environment and protection section of Institute of Arable Crops Research report, 1993.

8   ETSU, 'An assessment of renewable energy for the UK', HMSO, 1994.

9   B. Hague, 'Crops to energy', *REview*, issue 21, December 1993.

10 Rolf Olsson, 'A new concept for reed canary grass production and its combined processing to energy and pulp', paper 6 at PIRA's Silsoe conference, 23 March 1994.

11 L. Mason, 'Government plants for 650,000 ha of NVZs announced', *Farmers Weekly*, 20 May 1994.

12 See note 8.

13 J. Barshay, 'Ukraine shivers as energy dwindles', *Financial Times*, 23 January 1994.

14 J. Barshay and M. Kaminski, 'Ukraine to boost coal imports and exports', *Financial Times*, 17 August 1994.

15 *Power from Plants: the global implications of new technologies for electricity from biomass.* Report in the Energy and Environmental Programme, published by Earthscan, London, 1994.

16 C. Flavin and N. Lenssen, *Power Surge: Guide to the Coming Energy Revolution*, Worldwatch Environmental Alert series, published by W. W. Norton & Co., 1994.

17 Ralph Overend, 'Bioenergy guidelines in the USA,' from the workshop 'Wood fuel — the green debate', held at Hatfield, 18 October 1994, to be published by DTI/ETSU.

18 Personal communication from ETSU, 5 August 1994.

19 J. Kilpatrick, 'Sorting out the energy crops of the future', *Arable Farming*, June 1994, p. 37.

20 'British business and biomass power', Border Biofuels, from 'Wood Fuel — the green debate', Hatfield, 19 October 1994, to be published by DTI/ETSU

CHAPTER 15

Contact Addresses

ADAS: High Mowthorpe
Duggleby
Malton
North Yorkshire YO17 8BP
Tel: 01944 3646

ADAS: Arthur Rickwood
    Research Centre
Mepal
Ely
Cambridgeshire CB6 2BA
Tel: 01354 692531

ADAS: Pwllpeiran Centre
Aberystwyth
Dyfed SY23 4AB
Tel: 01974 22 229

Association of Independent
    Electricity Producers
1st Floor
41 Whitehall
London SW1A 2BX
Tel: 0171 930 9390

Avon Vegetation Research
PO Box 1033
Nailsea
Bristol BS19 2FH

Banks Doltons Ltd
Hermitage
Hampshire RG16 9QU
Tel: 01635 200733

Banks of Sandy Ltd
29 St Neots Road
Sandy
Bedfordshire SG19 1LD
Tel: 01767 680351

Alick Barnes
Loyton Farm
Morebath
Nr Tiverton
Devon EX16 9AS
Tel: 01398 331051

Border Biofuels Ltd
10 Currie Street
Duns
Berwickshire TD11 3DL
Tel: 01361 88363

British Biogen
22 Long Acre
London WC1E 9LY
Tel: 01435 882 228

Rupert Burr
Roves Farm
Sevenhampton
Swindon
Wiltshire SN6 7QG
Tel: 01793 763939

Murray Carter
Ingerthorpe Hall
Farm Markington
Harrogate
North Yorkshire HG3 2PD
Tel: 01765 677887

Centre for Agricultural Strategy
University of Reading
2 Earley Gate
Whiteknights Road
Reading RG6 2AU
Tel: 01734 861101

Council for the Protection of Rural
    England
Warwick House
25 Buckingham Palace Road
London SW1W OPP
Tel: 0171 976 6433

Country Landowners'
    Association
16 Belgrave Square
London SW1X 8PQ
Tel: 0171 235 0511

Edward de Lisle
Drayton Estate Office
Lowick
Kettering
Northamptonshire NN14 3BG
Tel: 01801 22405

Department of Agriculture,
    Northern Ireland
Dundonald House
Upper Newtownards Road
Belfast BT4 3SB
Tel: 01232 524619

East Midlands Electricity plc
Generation Division
Caythorpe Road
Caythorpe
Nottingham NG14 7EB
Tel: 01602 269711 or 620033

ESD Ltd
Overmoor Farm
Neston
Corsham
Wiltshire SN13 9TZ
Tel: 01225 812102

ETSU: see under Renewable
    Energy Enquiries Bureau

FEC EUS
Wellington House
63 Queen's Road
Oldham
Manchester OL8 2BA
Tel: 01616 525331

Forestry Commission
    (Publications)
231 Corstorphin Road
Edinburgh EH12 7AT
Tel: 0131 334 0303

Forestry Commission (Research)
Alice Holt Lodge
Wrecclesham
Farnham
Surrey GU10 4LH
Tel: 01420 22255

Forestry Contracting Association
Dalfring
Inverurie
Aberdeen AB51 5LA
Tel: 01467 651368

Friends of the Earth
Tel: 0171 490 1555

The Game Conservancy Trust
Fordingbridge
Hampshire SP6 1EF
Tel: 01425 652381 or 656713

Gannon UK Ltd
Welbourn
Lincoln LN5 0QL
Tel: 01400 72475

Robert Goodwin
Ashman Farm
Kelvedon
Essex CO5 9BT
Tel: 01376 573236

Greenpeace
Tel: 0171 354 5100

Hadlow College of Agriculture
    and Horticulture
Hadlow
Tonbridge
Kent TN11 0AL
Tel: 01732 850551

J. J. Harvey
Easterhill Gardens
Kenn
Exeter
Devon EX6 7UG
Tel: 01392 832693

Henry Doubleday Research
    Association
Ryton Organic Gardens
Ryton-on-Dunsmore
Coventry CV8 3LG
Tel: 01203 303517

Lionel Hill
Dunstall Court
Feckenham
Nr Redditch
Worcestershire B96 6QH
Tel: 01527 575751

Home Grown Energy Ltd
Maundrell House
The Green
Calne
Wiltshire SN11 8DL
Tel: 01249 821242

Hook Park College
Beaminster
Dorset DT8 3PH
Tel: 01308 863130

Dr Stig Ledin: see under Swedish
    University of Agricultural
    Science

Long Ashton Research Station
Long Ashton
Bristol BS18 9AF
Tel: 01275 392181

Martin Lishman
63 Pinchbeck Road
Spalding
Lincolnshire PE11 1QF
Tel: 01775 722464

LRZ Ltd
Unit 39 Park Farm Industrial
  Estate
Buntingford
Hertfordshire SG9 9AZ
Tel: 01763 272558

MAFF Market Task Force
Room 615
Nobel House
17 Smith Square
London SW1P 3JR
Tel: 0171 238 6600

MAFF
Whitehall Place
London SW1 2HH
Tel: 0171 270 8973

National Farmers Union of
  England and Wales
22 Long Acre
London WC2E 9LY
Tel: 0171 235 5077

National Farmers Union of
  Scotland
West Mains
Ingliston
Newbridge
Midlothian EH28 8LT
Tel: 0131 335 3111

National Westminster Bank
  Agricultural Office
24 Broadgate
Coventry CV1 1NB
Tel: 01203 553721

Northern Ireland Horticultural
  and Plant Breeding Station
Manor House
Loughgall
Co. Armagh BT61 8JA
Tel: 01762 891206

The Poplar Tree Company
Lower Ludlam
Madley
Hereford HR2 9JJ
Tel: 01981 250253

Rappa Fencing Ltd
Steepleton Hill
Stockbridge
Hampshire SO20 6JE
Tel: 01264 810665

Renewable Energy Enquiries
  Bureau
ETSU
Harwell
Oxfordshire OX11 0RA
Tel: 01235 432450

Royal Botanic Gardens
Kew
Richmond
Surrey TW9 3AB
Tel: 0181 940 1171

Royal Institute of International
  Affairs
10 St James's Square
London SW1Y 4LE
Tel: 0171 957 5736

Samuel Rose
The Lodge
Brixworth
Northamptonshire NN6 9BX
Tel: 01604 882255

Rosewarne
Camborne
Cornwall TR14 0AB
Tel: 01209 716674

Scottish Crop Research Institute
Invergowrie
Dundee DD2 5DA
Tel: 01382 562731

Silsoe College (Cranfield
   University)
Silsoe
Bedfordshire MK45 4DT
Tel: 01525 60428

Silsoe Research Institute
Wrest Park
Silsoe
Bedfordshire MK45 4HS
Tel: 01525 860000

Southern Electric PLC
Westacott Way
Littlewick Green
Maidenhead
Berkshire SL6 3QB
Tel: 01628 822166 or 01256 464044

South Western Power
800 Park Avenue
Aztec West
Almondsbury
Bristol BS12 4SE
Tel: 01454 201101

Edward Stenhouse Ltd
Newbridge
Coleman's Hatch
Hartfield
East Sussex TN7 4ES
Tel: 01342 826661

Swedish University of
   Agricultural Science
Department of Ecology and
   Environmental Research
   Section of Intensive Short
   Rotation Forestry
Box 7072
S-75007
Uppsala
Sweden
Tel: 00 46 186 72409

Talbott's Ltd
Drummond Road
Astonfields Industrial Estate
Stafford ST16 3HJ
Tel: 01785 213366

Timber Growers' Association
5 Dublin Street Lane South
Edinburgh EH1 3PX
Tel: 01317 7111

Voelund Danstoker AS
Nordist
The Broyle
Lewes
East Sussex BN8 6PH
Tel: 01825 841222

Water Research Centre plc
Henley Road
Medmenham
Marlow
Buckinghamshire SL7 2HD
Tel: 01491 571531

Christopher Whinney
Holdridge Farm
North Molton
North Devon EX36 3HG
Tel: 01598 4314

Wood Energy Development
  Group: see under Murray
  Carter

US National Renewable Energy
  Laboratory
PO Box 2008, Bldg 1505
Oak Ridge
Tennessee 3781–6038
USA
Tel: 001 615 576 7756

West Cornwall Leader Project:
  see under Rosewarne

Wood Fuel Research Group
School of Agriculture
581 King Street
Aberdeen AB9 1UD
Tel: 01224 272000

Index

Page numbers in italic indicate information is found only in the illustration or caption. On page numbers followed by 'tab', information is found only in the table.

206

FARMING PRESS BOOKS & VIDEOS

Below is a sample of the wide range of agricultural and veterinary books and videos we publish. For more information or for a free illustrated catalogue of all our publications please contact:

Farming Press Books & Videos, Wharfedale Road, Ipswich IP1 4LG, United Kingdom
Telephone (01473) 241122   Fax (01473) 240501

New Hedges for the Countryside   Murray Maclean

Gives full details of hedge establishment, cultivation and maintenance for wind protection, boundaries, livestock containment and landscape appearance.

Farm Woodland Management

Blyth, Evans, Mutch & Sidwell

Covers the full range of woodland size from hedgerow to plantation with the emphasis on economic benefits allied to conservation.

Farming and the Countryside

Mike Soper & Eric Carter

Traces the middle ground where farming and conservation meet in co-operation rather than confrontation.

British Farming: changing policies and production systems   Eric Carter & Malcolm Stansfield

An introduction to the main types of farming enterprise, their current changes and their wider implications – ideal for schools.

Organic Farming   Nicolas Lampkin

An outstanding wide-ranging account of the principles and practice for livestock and crops.

Farming Press Books & Videos is part of the Morgan-Grampian Farming Press Group which publishes a range of farming magazines: *Arable Farming, Dairy Farmer, Farming News, Pig Farming, What's New in Farming.* For a specimen copy of any of these please contact the address above.